NSF-CBMS Regional Conference Series
in Probability and Statistics
Volume 5

MIXTURE MODELS: THEORY, GEOMETRY AND APPLICATIONS

Bruce G. Lindsay
Pennsylvania State University

Institute of Mathematical Statistics
Hayward, California

American Statistical Association
Alexandria, Virginia

Conference Board of the Mathematical Sciences

Regional Conference Series in Probability and Statistics

Supported by the
National Science Foundation

The production of the *NSF-CBMS Regional Conference Series in Probability and Statistics* is managed by the Institute of Mathematical Statistics: John Collins, IMS Managing Editor/Statistics; Patrick Kelly, IMS Production Editor; Miriam Gasko Donoho, IMS Treasurer; and Barbara J. Lindeman, IMS Business Manager.

Library of Congress Catalog Card Number: 94-75429

International Standard Book Number: 0-940600-32-3

Printed in the United States of America

Acknowledgments

I have many persons to thank for their part in the writing of this monograph. First and foremost, John Grego was responsible for the successful proposal that led to the lecture series. He also assumed the onerous task of being the local host. Without him, and his strong supporting cast from the University of South Carolina, neither the lectures nor the notes would have come into being. Of course, I am also indebted to the NSF and CBMS for their sponsorship of the conference and the lecture notes.

Much of the work reported in this set of notes is based on collaborative efforts of myself with others. I owe much to the many people who I have worked with over the years on these problems. I must thank especially Kathryn Roeder and Dankmar Böhning, as a considerable bit of what I report here is due to them. While engaged in this research, I have been provided research support by the National Science Foundation and the Humboldt Foundation of Germany.

In addition, my current graduate students have all contributed in a substantial way to the final product you have in your hands. In the last frantic days of finishing, they pulled together with me to help construct the many figures, tables, and the bibliography. For this, I thank Matilde Sanchez, Liwen Xi and Ramani Pilla for their efforts, and add special thanks to Yanling Zuo for the bibliographical work and Yong Lin for the many figures found in Chapter 4.

I am also indebted to those friends and family members who contributed, both by positive support and by accepting my extended working hours. Finally, the book is dedicated to my parents, George Speers Lindsay (1907–1994) and Geneva Elizabeth Lindsay (1907–), whose formative role in my life becomes more and more apparent to me. The best that I have, I got from them. An additional dedication is to Cifford Clogg (1949–1995), whose premature death was a great personal loss to me and a tragedy for our profession.

Acknowledgments

Contents

CHAPTER 1

The Wide Scope

The mixture model has long been a challenge to the statistician, whether beginner, practitioner or theoretician. Recent times have seen great advances in our understanding of the some basic mathematical features of this model, and these notes are meant to be a unification of the work I have carried out, jointly with many wonderful collaborators, in this area. Based on lectures given in 1993 at a regional conference of the Conference Board of the Mathematical Sciences, the notes are directed toward a mixed audience of advanced graduate students and research workers in this and related areas. For the sake of newcomers to the mixture model, I will attempt to be complete enough for the text to make sense in itself, but must at some points refer the reader to other more extensive treatments.

Unfortunately, the goal of timeliness in the end also forced some truncation of the subject matter in the original lecture notes. On the other hand, some subjects have been given enhanced development because they are truly new, and the audience I have in mind will appreciate a deeper presentation of background and of the beautiful geometric structures of the model.

The first chapter of these notes, corresponding to Lecture 1, lays out the mixture landscape as I see it, and the practical side of my motivation for interest in the area. There are two major points:

- There are *many* statistical topics, some quite extensive by themselves, that can rightly be called mixture model subtopics. They all share the mixture model structure, and have similar inferential goals. One of the themes of this chapter is, therefore, that the many names for the mixture model hide its universality. In these notes I am aiming for the universal aspects that lie beneath.
- There is a nonparametric approach to maximum likelihood in the mixture model that gives us an extremely powerful set of tools to use, both in a nonparametric approach and in diagnosing the ailments of parametric approaches. I might add that, to my mind, it is mathematically elegant and fun.

The more sophisticated reader knows well that the sphere of mixture modeling includes random and mixed effects models, empirical Bayes, latent class and trait models, clustering, deconvolution and many other key words and phrases, and for such a person this chapter could perhaps be shorter. Just the same, it was an adventure and an amazement to me just to construct an itemized list.

My own understanding of this subject has been greatly enhanced by studying the geometric structures of the model, and as such the emphasis in later chapters will many times be on fundamental geometric ideas whose truth can sometimes be transmitted more easily pictorially than through detailed mathematical arguments. In addition to giving those pictures, I will sometimes sacrifice the generality of an argument in order to make it more transparent, with the understanding that the reader pursuing the matter further should go to the cited sources. The reason for this is that the *goal* in these notes is not to repeat what has been done, but to try to provide a reader of modest background a clear understanding of a sometimes difficult topic.

Although this is not a textbook, I am presuming that some readers will wish to confirm and enhance their understanding by the active process of doing calculations and drawing pictures rather than through the more passive activity of reading. For this purpose, I have marked various features of the text as exercises (denoted by *Italic* type and frequently, but not always, within brackets). None should take long to perform, given that the desired insight has been obtained.

1.1. The finite mixture problem. The simplest and most natural derivation of the mixture model arises when one samples from a population that consists of several homogeneous subpopulations, which we will call the *components* of the population. The number of components will generally be denoted m, but if we wish to emphasize that it is not known, we will use ν. The components will be indexed by $j = 1, \ldots, m$. Suppose we sample from such a population, recorded as data (X_i, J_i), for $i = 1, \ldots, n$, where $X_i = x_i$ is a measurement on the ith sampled unit and $J_i = j_i$ indicates the index number of the component to which the unit belongs.

Further, suppose that if we were sampling just from the jth component, it is known that there would be an appropriate probability model for the sampling distribution, say

$$\Pr[X = x | J = j] = f(x; \theta, \xi_j). \tag{1.1}$$

Please note that although the left side of the above expression formally refers to a discrete variable X, we will use this same symbol—and its like—in these notes to mean a density function, whether discrete or absolutely continuous. In (1.1), f represents a known density function, most naturally called the *component density*.

The variable ξ_j in (1.1) is an unknown parameter, called the *component parameter*, that describes the specific attributes of the jth component population, and θ is a vector of parameters that describes unknown characteristics common to the entire population. For the moment, the parameter θ is of secondary interest and will be dropped from the notation, although we will return to it later.

The proportion of the total population that is in the jth component will be denoted by π_j and called the *component weight*. We therefore have $\sum_j \pi_j = 1$. The component weights π_j are usually unknown parameters. Since we suppose that the population has been sampled at random, the probability that an observation comes from the jth component is $\Pr(J = j) = \pi_j$. We can conclude that the variables (X_i, J_i), for $i = 1, \ldots, n$, are a random sample from the joint density described by

$$\begin{aligned}\Pr(X = x, J = j) &= \Pr(X = x|J = j)\Pr(J = j)\\ &= f(x;\xi_j)\cdot\pi_j.\end{aligned}$$

The mixture model arises if the component label data $J_1, \ldots, J_n$ is missing, so that we observe only the sample $X_1, \ldots, X_n$ from the marginal density of X. Thus the observed data are a sample from the *mixture* density

$$g(x;\boldsymbol{\pi},\boldsymbol{\phi}) = \sum P(X = x|J = j)P(J = j) = \sum \pi_j f(x;\xi_j). \tag{1.2}$$

If there are m components in the mixture, it will be called an m-component *finite mixture model*. In a general finite mixture model, there is a set of $2m$ parameters

$$\begin{pmatrix}\pi_1 & \cdots & \pi_m\\ \xi_1 & \cdots & \xi_m\end{pmatrix},$$

each column corresponding to a component. The weights satisfy the constraints that

$$\pi_j \geq 0 \quad \text{and} \quad \sum_j \pi_j = 1.$$

A very important special case occurs when there is just one component; in this case the density $f(x;\xi)$ will be the called the *unicomponent* mixture density.

1.1.1. *A simple example.* To illustrate some of the fundamental characteristics of the problem, it is useful to consider a simple example and examine the mixture densities that arise. Suppose that we have a population of animals consisting of two component types, say component 1 = male and component 2 = female, and the characteristic, say X = length, is normally distributed in both components when considered alone. Suppose for simplicity that the two component groups have the same standard deviation σ for X, but that they have different means, ξ_1 and ξ_2, respectively. Further, assume that the males have the smaller mean ξ_1, so $\xi_1 < \xi_2$ (e.g., black widow spiders). Let π be the population proportion of component 1 so that $\bar{\pi} := 1 - \pi$ is the proportion from component 2. If we sample from the two components without gender label, then the resulting distribution for heights is a mixture of two normals,

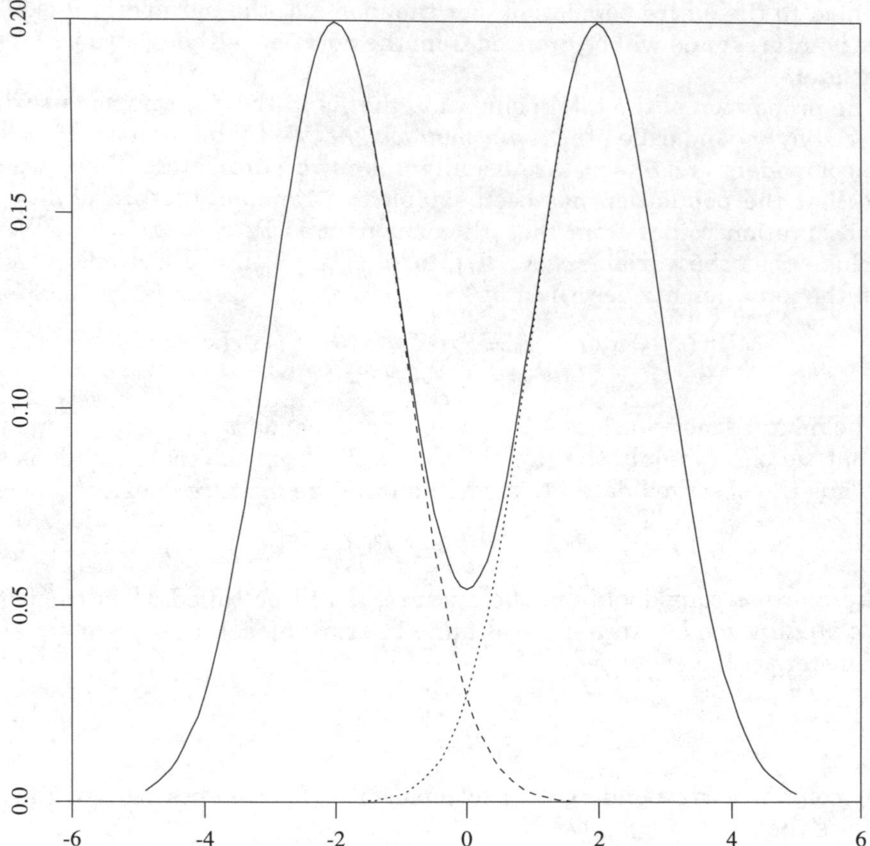

FIG. 1.1. *Mixed normal densities with means four standard deviations apart.*

which we write as

$$\pi N(\xi_1, \sigma^2) + \bar{\pi} N(\xi_2, \sigma^2).$$

Here we use the symbol $N(\xi, \sigma^2)$ to represent the normal probability measure with mean ξ and variance σ^2.

If the components have equal component weights of $\pi = (1 - \pi) = 1/2$ and the means are four standard deviations apart, then heights from the population are described by the solid curve in Figure 1.1.

In this case the mixed nature of the density is revealed through its bimodality. Moreover, it is clear that we can obtain from the value of the variable X = height a great deal of information about whether the measurement was on a male or female.

We might contrast this favorable situation with one in which the means are just one-half as far apart, two standard deviations. One might not expect the dramatic change that going from four to two standard deviations makes. In Figure 1.2 we see that the population density is now unimodal.

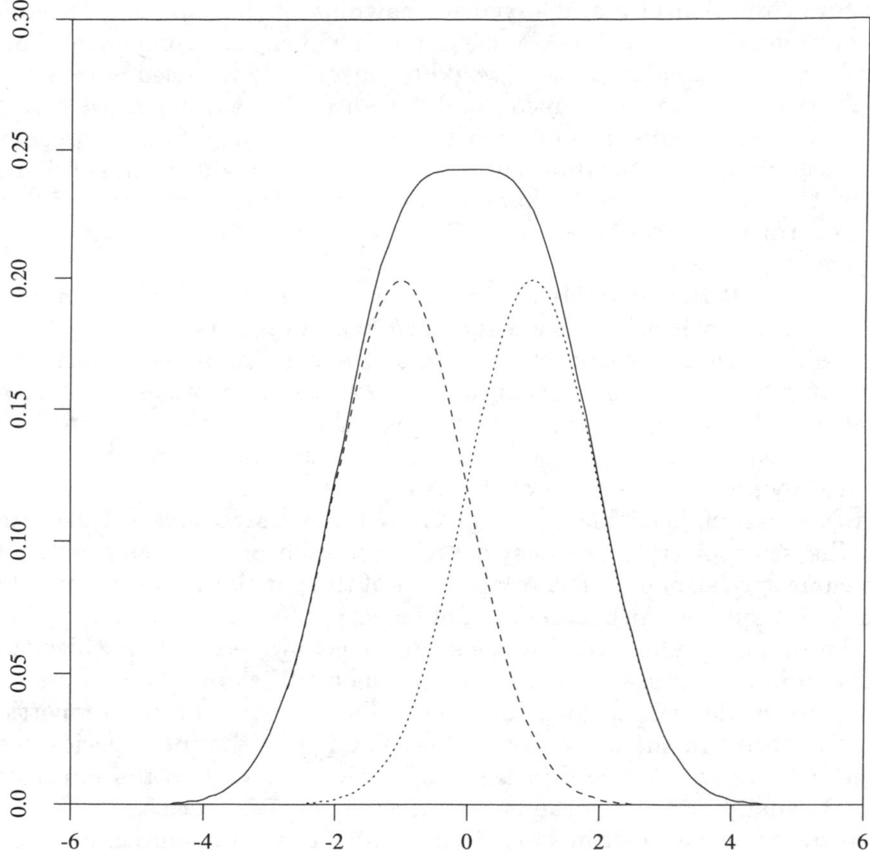

FIG. 1.2. *Mixed normal densities with means two standard deviations apart.*

Suppose that we wished to construct a classification rule that took the variable X and made a determination of the sex. In this case, the best rule (fewest misclassifications) would be to assign those spiders smaller than the population mean to be males and those larger than this to be females. As an indication of the loss of information about the true number of males and females, we note that using this rule would cause many males and females to be misclassified. (We will learn more about the difficulties associated with the loss of information about the mixture due to the closeness of the components in a later section of this chapter.)

1.1.2. *More complicated applications.* The most compelling examples of the mixture model occur when there are physically identifiable components in a true population. We would like to point out here just two of the many interesting such applications in the literature.

MacDonald and Pitcher (1979) considered a situation in which the population consists of the fish of a single species in a lake, and each component consists of the fish of a single yearly spawning of that species. The components are thus relatively homogeneous and fish in any one component j might, therefore, have $X =$ length that would be adequately modeled with a normal distribution with unknown mean and variance. The parameters ξ_j are then the parameters in this normal distribution and the weights π_j represent the relative abundance of the different age groups. The weights would often be the parameters of interest. Note that determining the values of these parameters over several years would be useful for determining the relative mortality of the yearly cohorts.

We note that this example illustrates one of the useful features of mixture analysis. It enables one to use a *surrogate* measure, in this case $X =$ length, in place of an ideal measurement, here $J =$ age group, in experimental situations in which the ideal measurement is expensive or impossible to obtain. However, in MacDonald and Pitcher's case, this ideal measurement is actually available, so they can compare their mixture model analysis with that attained by knowing the ages of the fish.

Another example of this type is given by Do and McLachlan (1984). In this case the scientific interest was in the population of rats being eaten by a particular species of owl. The components of the population were a number of distinct rat species. Although it would be very difficult to directly survey the owls for dietary preference, it was easy to collect owl feces, from which the rat skulls could be extracted. Various measurements X were taken on the skulls. A mixture model can be constructed as follows: Let π_j be the proportion of the jth species in the owls' diet and let $f(x;\xi_j)$ be the intraspecies density of these characteristics within the jth species, where f is the multivariate normal density and ξ_j are the parameters for the jth species.

For the reader who wishes to read more about the many direct applications of the finite mixture model, there are three good books available on the topic: Everitt and Hand (1981), Titterington, Smith and Makov (1985) and McLachlan and Basford (1988). We return to the basic mathematical structures of the model.

1.2. The latent (or mixing) distribution. An extremely important aspect of the finite mixture problem that we have just described is that we can identify the unknown parameters with a distribution. We next show how this is done.

1.2.1. *The discrete latent distribution.* Define the *latent* random variables $\Phi_1, \ldots, \Phi_n$ to be the values of the parameter ξ corresponding to the sampled components $J_1, \ldots, J_n$; that is, if the ith observation came from the jth component, then define $\Phi_i = \xi_j$. In symbols,

$$\{J_i = j\} \iff \{\Phi_i = \xi_j\}.$$

It is conventional to let the realized value of a random variable, such as Φ_i, be denoted by the lowercase version of the same letter, such as $\Phi_i = \phi_i$. Thus we will have two symbols, ϕ and ξ, representing elements of the component parameter space. It is useful to do so in order to avoid ambiguity about whether a symbol, such as ϕ_j, refers to the component parameter of the *jth component* (here denoted ξ_j) or the latent variable that was sampled in the *jth observation* (here denoted ϕ_j).

The Φ_i are a random sample from the discrete probability measure Q that puts mass π_j at the support point ξ_j; that is,

$$\Pr[\Phi = \xi_j] = Q(\{\xi_j\}) = \pi_j.$$

We can in this way equate the set of unknown parameters π_j and ξ_j uniquely with a discrete probability measure Q on the parameter space for ξ, with m points of support $\{\xi_1, \ldots, \xi_m\}$ and corresponding masses $\{\pi_1, \ldots, \pi_m\}$. Thus we have the extremely important concept that estimating the unknown parameters in the mixture model (1.2) is the same as determining an unknown distribution Q with support on the parameter space.

The distribution Q is usually called the *mixing distribution*, but to avoid confusion with the expression "mixture distribution" and because it is associated with the latent variable Φ, here we will call it the *latent distribution*, in line with terminology found in the social sciences literature. We will also replace the component and weight parameters in the notation with Q, so that when it is known that we are in a finite mixture situation,

$$Q \equiv \begin{pmatrix} \pi_1 & \cdots & \pi_m \\ \xi_1 & \cdots & \xi_m \end{pmatrix}.$$

With this change in perspective, there is also a natural change in terminology, in which the component parameters ξ_j are now the *support points* of the latent distribution and the component weights π_j are the *probability masses* of the latent distribution.

We note that we can also view the hypothetical complete data set as being (X_i, Φ_i), $i = 1, \ldots, n$, rather than (X_i, J_i). Since we arrive at the same marginal distribution for the observable variables X_i, we can choose the representation that is most useful. The latent variable representation (X_i, Φ_i) links the mixture problem to the standard Bayesian paradigm: The mathematical model for the pair (X, Φ) is identical to a Bayesian model in which the conditional distribution of X given the realized parameter $\Phi = \phi$ is $f(x|\phi) = f(x; \phi)$, and Q is the prior distribution on Φ.

The distinction here is that we have more than one observation of the random variable X, corresponding to multiple realizations of the latent variable Φ, so that we are able to do frequentist inference about the prior Q itself. We return to this point later in discussing empirical Bayes. We note that a Bayesian approach to the mixture problem we are considering here can be found in the literature under the key phrase *hierarchical Bayes*.

We next note that if the component density f does not depend on the component other than through the parameter ξ, then we can write the mixture density of an observation X as an expectation, or average, over the latent distribution Q of Φ :

$$f(x;Q) := E[f(x;\Phi)] = \int f(x;\phi)\,dQ(\phi).$$

[*Exercise.*] This last representation of the mixture density as an integral of a known component density function f with respect to an unknown probability measure Q will be the basis of many of the results of this monograph.

1.2.2. *The continuous latent distribution.* Although we have derived the mixture model in the context of a population model with finitely many distinct components because it provides a fundamental level of understanding, it is important that there is a natural extension of the model in which the latent variable Φ has a continuous density, say $dQ(\phi) = q(\phi)\,d\phi$, so that the mixture density becomes

$$f(x;Q) := \int f(x;\phi)\,dQ(\phi) = \int f(x;\phi)\,q(\phi)\,d\phi.$$

Sometimes the continuous latent variable has a direct physical interpretation, in the sense that one could, at least in theory, measure it exactly, but that it was not measured on the ith unit. An example could be a variable such as $\Phi =$ age in a population with year-round births. The density $f(x;\phi)$ then represents the conditional distribution $f(x|\phi)$ of X given the missing variable $\Phi = \phi$. In this situation, it is possible to have a subsample of data in which both Φ and X are measured, with joint density $f(x;\phi)q(\phi)$.

Many other times Φ represents a more abstract quantity presumed to have a strong influence on the measurement X, and about which inference may be desired, but which cannot itself be directly determined by physical measurement. For example, we might suppose that there is a latent *mathematical ability* that largely determines the test score in mathematics on a certain exam, such as the SAT, but we realize that there is randomness in the outcome of an exam for any one subject arising from a variety of other factors.

We can allow for this by constructing a model in which the latent variable Φ_i is a subject-specific parameter that determines the probability that the ith subject will get a question right. Then each subject has a random number of correct answers, but his overall distribution is determined by that subject's latent *ability* parameter ϕ_i. Since the true value of this parameter can only be exactly determined through the taking of infinitely many test items of the same type, it has only abstract meaning.

Much of our interest in these notes will be in the case where the distribution Q is treated as completely unspecified as to whether it is discrete, continuous or in any particular family of distributions; this will be called the *nonparametric mixture model*. If the parameter θ is present, it will be called the *semiparametric mixture model*. In addition, it should be noted that we will later consider further complications to these models, such as the presence of covariates.

We note that these models have many applications, but using the nonparametric model is not appropriate when, for physical or other reasons, there is a discrete mixing distribution with a known number of components. However, the insights gained into the models by the nonparametric approach will be useful in this case as well, as we intend to demonstrate.

1.3. Many more variations and names. In this section we wish to make a brief compendium of the many statistical problems that have mixture structure, in the sense that there is an unknown probability distribution Q that enters naturally into the model construction. The mixture structure goes by many names and the main objective here is simply to alert the reader to the large number of application areas for the methodology discussed here. Many of the subjects mentioned here have vast literatures of their own, which we will not try to summarize. However, they all share the following feature: The likelihood on a observation can be written in the form

$$L_i(\theta, Q) = \int L_i(\theta, \phi)\, dQ(\phi),$$

where some features of the distribution Q are unknown and to be inferred from the data. The term $L_i(\theta, \phi)$ will be called the *likelihood kernel* and it represents the form of the density for the ith observation, conditional on $\Phi = \phi$.

1.3.1. *Known component densities.* There are many interesting and important examples in which there are a finite set of component densities that are completely known and so there are no unknown component parameters ξ_j. In this case we can retain the formal structure defined above by equating the latent variable Φ with the component variable J. The known density function for X conditional on $J = j$ can then be written as

$$\Pr[X = x | J = j] = f(x; j)$$

and the only unknown parameters are the component weights π_j. In this case, the latent distribution Q is the distribution of the variable J, and so knowing the component distributions is equivalent to specifying that the latent distribution Q has a known set of support points $\{1, 2, \ldots, m\}$. This case, called hereafter the *known component density model*, is simpler than the general problem, but is worth studying in its own right both for its many applications and for the insight it gives into the general problem. We therefore offer several examples of its use.

This author had his earliest exposure to the mixture problem in a consulting problem involving fish stock analysis. The components involved were salmon subpopulations, each identifiable as spawning in a single river system and possessing, therefore, a great amount of genetic homogeneity relative to the entire population. The measured variable X on each fish was a genetic typing determined by electrophoresis, a variable that had only a finite number of outcomes. Correspondingly, the jth river system would have a discrete density

$$f(t; j) = P(X = t|J = j)$$

describing the distribution in that river of the finite set of genetic types t. The data of interest arose when salmon were caught in a particular region of open ocean by fishermen and it was desired to know the fractions π_j of the region's salmon population that came from each river system. The research had political implications in that the river systems were in two countries, the United States and Canada. [For more on applications of this type, see Millar (1987).]

A second example of the known component type comes from Roeder, Devlin and Lindsay (1989). The data consisted of the genetic types of seeds that had been collected from a mother plant and it was desired to know what fraction π_j of the seed generation could be ascribed to each of several competing father plants. In this case, knowing the jth father's genetic type and the mother's genetic type, it was possible to construction a probability density $f(t; j)$ for the genetic types of the seed generation.

A third example arises in the use of positron emission tomography. In this problem, a subject ingests a radioactive substance which is designed to congregate in some bodily part of interest, say a tumor in the brain. An array of sensors is set up around the region of interest and the radioactive emissions are observed. The emissions have the following characterization: at the time of disintegration, two rays shoot off in opposite directions (180° apart). When they arrive at two opposing detectors, it can be deduced that the emission occurred at some unknown point on the line between the two detectors, but nothing more. Since the array has a finite number of detectors, there are a finite number of such opposing detector pairs. The index t will refer to the possible detector pairs and each emission generates an observed variable X taking on values in this set.

We desire to find the hot spots in the brain that are generating the most emissions. To do so, we create a grid of possible emission regions. The variable j will index this grid of sites in the brain, and we define a latent variable by letting $J_i = j$ mean that the ith emission came from site j. Thus we are interested in inference on $\pi_j = \Pr(J = j)$. The geometry of the array and the nature of the emissions determine exactly the component density $P(X = t|J = j) = f(t; j)$, the probability that an emission from j will be observed in the detector pair t.

Thus we have a simple mixture model for the observed process of emissions: $P(X = t) = \sum \pi_j f(t; j)$. For more details, see Shepp and Vardi (1982) and Vardi, Shepp and Kaufman (1985).

1.3.2. *Linear inverse problems.* We offer yet another perspective on the mixture problem. Our basic unknown is the latent distribution Q that satisfies the relationship

$$g(x) = \int f(x; \phi)\, dQ(\phi). \tag{1.3}$$

Here f is known, but g is observed with some error— in our case because we see a sample from it. In the mathematics literature, solving for Q in (1.3) is called a *Fredholm integral equation of the first kind* [Wing and Zahrt (1991)].

If the density for X is discrete, with finite range $1, 2, \ldots, T$, and Q is discrete on a known support set $\{1, \ldots, m\}$, as in the examples of the preceding section, then we can write (1.3) as $g(t) = \sum_s f(t; s)\pi(s)$. This can be written as a matrix equation of the form

$$\mathbf{g} = \mathbf{F}\boldsymbol{\pi},$$

in which we wish to solve for the vector $\boldsymbol{\pi}$, where $\sum \pi_j = 1$ and the π_j are nonnegative. This, in the terminology of Vardi and Lee (1993), is a *linear inverse problem* with *positivity constraints*.

A recurring theme of the mathematical literature is that the linear inverse problem is difficult and unstable to solve. In the matrix case, this is clearly related to the numerical instability of solving such linear equations when the column vectors are highly correlated. One of the key points of Vardi and Lee is that the Expectation-Maximization (EM) algorithm (Chapter 3) is a stable and reliable way to handle the numerical difficulties involved in solving linear inverse problems.

1.3.3. *Random effects models.* Consider the classic one-way ANOVA model. We can visualize the data as being in a two-way array:

$$\begin{array}{cccc} X_{11} & X_{12} & \cdots & X_{1n_1} \\ X_{21} & X_{22} & \cdots & X_{2n_2} \\ \vdots & \vdots & \vdots & \vdots \\ X_{p1} & X_{p2} & \cdots & X_{pn_p} \end{array}.$$

In the *fixed effects* version of this model, row i consists of a sequence of i.i.d. observations from a $N(\phi_i, \sigma^2)$ density, where the parameters $\phi_1, \ldots, \phi_p$ are the unknown means of the p rows. In the *random effects* version of this model, the parameters are viewed as having been sampled from a larger population and so have a latent distribution Q in that larger population. The resulting marginal density of the observations in a single row is the mixture model

$$f(\mathbf{x}; Q, \sigma^2) = \int f(x_1; \phi, \sigma^2) \cdots f(x_n; \phi, \sigma^2)\, dQ(\phi).$$

The usual assumption in the normal theory random effects model is that Q is a normal distribution. If we make this assumption, then the observations $\mathbf{X}$ in a row have a multivariate normal distribution, with a positive and equal correlation between observations that is induced by the fact that the entire row has the latent variable ϕ in common. In fact, the covariance matrix for the row vector $\mathbf{X}$ has the form $\sigma^2 I + \sigma_\Phi^2 E$, where I is the identity matrix and E is a matrix of ones. [*Exercise*: This is easiest to derive by writing $\mathbf{X} = \mathbf{Z} + \Phi \cdot \mathbf{1}$.]

Although the assumption of the normality of the latent row parameters may be tenable in many examples and is the simplest path, it is possible to take a nonparametric approach to the problem. This latter approach seems especially natural when the data are discrete so that normality of the observations is no longer a tenable assumption. We will consider some aspects of an example of this type in Chapter 8.

1.3.4. *Repeated measures models.* Another statistical problem with nearly the same structure as the random effects model is the *repeated measures model*. If we think of the row vector $\mathbf{X}$ as being a sequence of repeated measurements on a single individual and assume that $\mathbf{X}_i$ has a multivariate normal distribution with

$$\text{Var}(X_{ia}) = \text{Var}(X_{ib}) = \sigma^2$$

and $\text{Corr}(X_{ia}, X_{ib}) = \rho$, then one has the exact same multivariate normal distribution for the observable vectors of $\mathbf{X}$ as in the normal random effects model above, with one exception: In the random effects model, ρ must be nonnegative.

We note that these models start from different modeling points of view, but arrive at essentially the same model. The random effects model arises from assuming the existence of subject-specific latent means, conditional upon which the individual observations within a subject are independent. The marginal correlation of observations within a subject is a consequence of that structure. In the repeated measures model, we have directly modeled the whole vector of observations, including its correlation structure. As a bonus, modeling the covariance structure directly allows us to consider dependencies other than the equicorrelation of all pairs, which, for example, would not be most natural if the repeated observations had occurred in a time series.

These different points of view are largely reconciled in the normal model, at least for equicorrelation, but when the normality assumption is not tenable, they have led to two different schools of modeling data. If we say that the observations in a row form a *cluster* because of their related nature, having all come from the same subject or school or other unit, then we expect them to have some correlation, most likely positive, when viewed from the perspective of the entire population. That is, the population covariance matrix of the $\mathbf{X}$ vector should show nonnegligible correlations. We can choose to model this directly, and do so in the so-called *population average* approach. The other model building approach is *cluster specific*. That is, we can model the effect of being in a cluster (row) by a cluster-specific latent parameter ϕ and assume that this parameter has been sampled from a population Q. This then induces a correlation structure.

When the observations have a binary nature, these two approaches to modeling result in rather different models. See, for example, Neuhaus, Kalbfleisch and Hauck (1991).

1.3.5. *Latent class and latent trait models.* As noted earlier, we have borrowed the terminology "latent variable" from the social sciences literature, where the expression is often used in the mixture model analysis of categorical data. In this literature, we have multinomial observations $\mathbf{X}_i$ whose probabilistic behavior is determined by some unobserved variable Φ_i. If the distribution of Φ is discrete, then the possible values of Φ (the support points) correspond to the *latent classes* of the population. If the variable is continuous, then its values correspond to some *latent trait* of the population. For a review of this literature, see Clogg (1995) and Heinen (1993).

1.3.6. *Missing covariates and data.* Whenever variates are missing at random from a data set, whether covariates or response variables, then the distribution of the remaining values comes from integrating the missing values out of the joint density function. If we have a regression model, say $f(y|x,z) = f(y|x'\beta + z'\gamma)$, where the z's are entirely missing, then we can write the conditional model for the observable data as $f(y|x) = \int f(y|x'b + \phi)\, dQ(\phi|x)$, where Q is the conditional distribution of $z'\gamma$. If the measured covariates are independent of the missing ones, so that $dQ(\phi|x) = dQ(\phi)$, then the resulting model is of mixture form, with a random intercept in the regression. That is, using a random intercept in a regression model is a method that allows for the additional uncertainty in the inference due to missing covariates.

1.3.7. *Random coefficient regression models.* Another area which has seen much work of a nonparametric kind is the random coefficient regression model. The previous examples in this section have largely been of a kind where a regression model for a particular cluster might include a random intercept term, as in $\beta' z + \phi$. There exist a number of studies in the pharmacokinetics literature of models in which one or more of the β's are random as well. See, for example, Mallet (1986), Mallet, Mentre, Steimer and Lokiec (1988) and Davidian and Gallant (1992). We also note the extensive study of this model by Beran and Hall (1992).

1.3.8. *Empirical and hierarchical Bayes.* Yet another application of the mixture methodology arises in the subject known as *empirical Bayes*. The basic conceptual framework is as we have described, but now the emphasis is on making inferences on the set of realized values of the latent variable $\Phi_1 = \phi_1, \ldots, \Phi_n = \phi_n$. If the distribution Q had been known in advance, then we are in a Bayesian setting and so the best mean-squared error estimator of an individual ϕ is the Bayes estimator

$$E[\Phi|X = x] = \frac{\int \phi f(x;\phi)\, dQ(\phi)}{\int f(x;\phi)\, dQ(\phi)}.$$

Since Q is unknown in our case, one natural tactic is to replace the Q in the above formula with an estimate, either parametric or nonparametric. However, there are many other strategies; see the book by Maritz and Lwin (1989).

One important aspect of empirical Bayes methods is that it has been learned that these methods have advantages in estimating a set of parameters $\phi_1, \ldots, \phi_n$ even if they did not arise as a sample from a distribution Q. To be more precise, the comparison we wish to make is between the fixed effects model, in which the ϕ parameters are treated as unknown parameters and so the likelihood has the form

$$L_{\mathrm{FE}}(\phi_1, \ldots, \phi_n) = \prod_i f(x_i; \phi_i),$$

and the random effects model, which has a likelihood of the form

$$\prod_i \int f(x_i; \phi)\, dQ(\phi) = \int \cdots \int L_{\mathrm{FE}}(\phi_1, \ldots, \phi_n)\, dQ(\phi_1) \cdots dQ(\phi_n).$$

The claim is that there are advantages to using the random effects model, even if conceptually the fixed effects model is more applicable, when one is estimating the values of the realized sequence of parameter values. For a readable introduction to this subject, the *Scientific American* article of Efron and Morris (1977) is highly recommended.

The rough idea is that if one estimates the parameters individually, say by $\hat{\phi}_i$, then the set of estimates $\{\hat{\phi}_1, \ldots, \hat{\phi}_n\}$ tends to be more dispersed than the set of parameters being estimated. To take an extreme case, if the true ϕ's are all equal, then the actual parameters have *no* dispersion, but the estimates have a dispersion corresponding to the variance of their distribution.

Bayesian formulations of this problem fall under the keyword *hierarchical Bayes*. In its simplest form, we say Q has some simple parametric form, say normal, and put priors on the parameters of this density.

1.3.9. *Nuisance parameter models.* The "empirical Bayes effect" just described also arises when estimating a nonlatent parameter of interest θ in the presence of many nuisance parameters $\phi_1, \ldots, \phi_n$, the famous Neyman–Scott problem (1948). This is a topic which has ended up being truncated from the notes, but not due to lack of importance.

The contrast is between the fixed effects approach, which treats the ϕ values as unknown parameters, and random effects approach, which treats them as an aggregate by assuming they come from an unknown distribution. We have already noted that there are efficiency advantages to the empirical Bayes approach when making inferences about the parameters ϕ_i. It turns out that there also are advantages in estimating the structural parameters θ.

1.3.10. *Measurement error models.* Suppose that our goal is to make scientific inferences about the relationship between two population variables, say $Y :=$ blood cholesterol level and $\Phi :=$ dietary cholesterol level. For example, we may wish to fit some sort of regression model, with parameter β, to the conditional density $f(y|\Phi; \beta)$. However, we have available only a *surrogate* measure X that is highly correlated with Φ, such as

$$X := \text{estimate of } \Phi \text{ from dietary questionnaire.}$$

It is well known that if we use X in place of Φ, then the resulting regression will show a diminished effect of the covariate X as compared with the use of latent variable Φ. Indeed, in a worst case scenario, where one covariate is measured with error and a second one is not, the relative importance of the two variables can be reversed.

One easy way to illustrate this point is to assume that (Y, Φ) have zero means and covariance matrix

$$\begin{bmatrix} \sigma_Y^2 & \rho\sigma_Y\sigma_\Phi \\ \rho\sigma_Y\sigma_\Phi & \sigma_\Phi^2 \end{bmatrix}.$$

In such a setting, the correct (latent) regression slope is $\beta = \rho\sigma_Y/\sigma_\Phi$.

Assume the surrogate satisfies $X = \Phi + \varepsilon$, where the measurement error ε is independent of the other variables. Since $\mathrm{Cov}(X, Y) = \mathrm{Cov}(\Phi, Y)$, the covariance matrix for (Y, X) differs from the above only in the lower right corner, where we obtain $\mathrm{Var}(X) = \sigma_\Phi^2 + \sigma_\varepsilon^2$, and so the surrogate (Y, X) regression coefficient is

$$b = \frac{\mathrm{Cov}(X, Y)}{\mathrm{Var}(X)} = \beta\frac{\sigma_\Phi^2}{\sigma_\Phi^2 + \sigma_\varepsilon^2},$$

so that $0 \le b/\beta \le 1$.

There are two approaches to the measurement error problem, corresponding closely to whether we treat the unobserved $\Phi_1, \ldots, \Phi_n$ as unknown parameters (by conditioning on their realized values $\phi_1, \ldots, \phi_n$) or as latent random variables from some unknown population distribution Q with its own set of unknown latent parameters. The former approach runs into the problem of infinitely many nuisance parameters mentioned earlier. Taking the latent variable or mixture approach, we write the density of the observed variables as

$$f(y, x) = \int f(y, x|\phi)\, dQ(\phi) = \int f(y|\phi, x) f(x|\phi)\, dQ(\phi).$$

Assuming that in the knowledge of ϕ, the variable X contributes no further information about Y, we replace $f(y|\phi, x)$ with $f(y|\phi)$, the regression distribution of interest. The density

$$f(x|\phi) = \Pr(X = x|\Phi = \phi)$$

is determined by the measurement error process and, typically, as in the above example, X would be modeled as having a normal distribution with mean $\Phi = \phi$. The important feature to us is that the mixture model structure holds in this case, with likelihood kernel

$$L_i(\phi) = f(y_i|\phi, x_i) f(x_i|\phi).$$

In Chapter 8 we will return to this model to analyze some properties of the nonparametric approach to estimating the distribution Q.

1.3.11. *Deconvolution problems.* Deconvolution problems are mixture problems with a special additional piece of structure. If the data X can be written as $\Phi + Z$, where Φ is a latent variable and Z has a known density f, then the resulting density for X has the mixture form

$$g(x; Q) = \int f(x - \phi)\, dQ(\phi).$$

This structure, as we have just seen, arises as a natural model for measurement error processes. Because of this structure, one can employ special techniques to attempt the inversion problem. In particular, because of the identity of characteristic functions, $\Psi_x(t) = \Psi_\Phi(t)\Psi_z(t)$, one can develop methods to solve for the characteristic function of latent variable Φ from the known characteristic function of Z and an estimated characteristic function of X.

Since the cumulant generating functions in a convolution are additive, one also has the cumulant identity $\kappa_r(X) = \kappa_r(\Phi) + \kappa_r(Z)$. If Z is normal, then this implies that the cumulants of X and Φ are identical for $r \geq 3$, since $\kappa_r(Z) = 0$ on this range. This simple structure has led to substantial use of moment methods in the signal processing literature, yet another statistics topic closely related to the mixture topic.

1.3.12. *Robustness and contamination models.* It is a common statistical practice to study the robustness of a statistical procedure by constructing a simple class of alternative mixture models. One can construct a simple symmetric model alternative to the normal model via the normal scale mixture,

$$(1-\alpha)N(\theta,\sigma^2) + \alpha N(\theta, k\sigma^2),$$

where k is large and α is small, representing the proportion of observations being measured with larger errors. In fact, many families of symmetric distributions that are commonly used as heavy tailed alternatives to the normal have *normal scale mixture* structure. For example, the family of t distributions are scale mixtures of normals. (*Hint:* Write $t = \Phi Z$, where Z is normal and Φ is independently distributed as the inverse of the square root of a chi-square divided by its degrees of freedom.)

Another approach to constructing robust procedures through mixtures are *contamination models*, such as $(1-\alpha)N(\theta,\sigma^2) + \alpha N(\phi,\tau^2)$. If α is small, ϕ quite different from θ and τ small, then this model generates *outliers* near the value ϕ.

Aitkin and Wilson (1980) explicitly used maximum likelihood with the above mixture models in order to obtain robust procedures, with apparent success. There appears to be little theoretical work on this approach to robustness.

1.3.13. *Overdispersion and heterogeneity.* The last section suggests that we might find additional advantages for the mixture type model in cases where there is no physical meaning to the latent variable Φ. It is just a simple means of extending our class of models to allow for any lack of fit of a basic model. As we will study in Chapter 2, the construction of a mixture model from a basic model will always give us not only flexibility, but a class of models that allows for more dispersion than the original model. This approach has long been used to construct models that allow for extra dispersion. For example, the beta binomial model, used as an extension of the binomial model, or the negative binomial model, used as an extension of the Poisson, both have natural derivations as mixtures of the basic models. These points will be discussed briefly in Section 3.1.1, which deals with conjugate families of distributions.

1.3.14. *Hidden mixture structures.* There are a variety of nonparametric statistical models with hidden mixture structure. By this we mean, first of all, that the class of distribution functions in the model is *convex*. That is, if F and G are in the model, then so is $(1-\alpha)F+\alpha G$. Second, there is a representation of all elements of the model as convex combinations of some basic class of distributions, say $\{F_\phi\colon \phi \in \Omega\}$. If so, we can write an arbitrary element of the model as $\int F_\phi \, dQ(\phi)$. (This basic class should be in the set of extreme points of the convex class if we seek identifiability for Q.)

The simplest example is the class of all distribution functions. If we let F_ϕ be the degenerate distribution at ϕ and let Q be the latent distribution, then Q is also the mixture distribution. We here have the equality of the latent variable Φ with the observed variable X.

A more sophisticated example concerns the class of distribution functions with nonincreasing density functions on $[0,\infty)$. This class is clearly convex and we seek a class of basic distributions F_ϕ. The solution is to let F_ϕ be the uniform distribution on $[0,\phi]$. [*Exercise.*] Here the latent distribution appears to have no intrinsic statistical meaning.

1.3.15. *Clustering: A second kind.* We have already pointed out one of the mixture-related uses of the word cluster. However, mixture methods are also prominently used in another area that is, technically, nonstatistical. Suppose we have a set of vectors $\mathbf{X}_1,\ldots,\mathbf{X}_n$, not obtained by sampling. They might represent a vector of numerical characteristics for a set of n species. We desire to find those species which are most similar to each other in these characteristics, forming thereby *clusters* of species. One of the approaches to such problems is to treat the data as if it were a sample from a mixture distribution. For example, if we desire two clusters, we might construct a reasonable two component mixture model, such as a mixture of two multivariate normal densities with mean vectors ϕ_1 and ϕ_2 and common covariance matrix Σ. After fitting the mixture, say by maximum likelihood, one can assign a data point to cluster 1 or cluster 2 depending on its posterior probability of being from that component. That is, we calculate

$$\Pr(J=1|X=x)=\frac{\pi_1 f_1(x)}{[\pi_1 f_1(x)+\pi_2 f_2(x)]}$$

and assign the observation to cluster 1 if and only if this conditional probability is greater than 0.5. This approach to *cluster analysis* is extensively discussed in the book by McLachlan and Basford (1988).

1.4. Be aware of limitations. We hope that the reader is now convinced that there are a multitude of interesting statistical applications that involve the estimation of an unknown distribution function Q and whose likelihoods have the formal structure of mixture problem. However, we must also confess that inference is a very difficult task, as we take observations from Q only indirectly. This section contains some warnings about the limitations of our procedures and our knowledge.

1.4.1. *Robustness characteristics.* It is natural and desirable to ask the question: What are the consequences of slight errors in the specification of the model? When a mixture model is being specified, and I specify a mixture of two normals with different means, what are the consequences if the mixture is actually of two t distributions or is actually a mixture of three normals, two of which are very close together? How stable are my parameter estimates under contamination? At present we would seem to know very little about this side of the subject. Most of the robustness literature is not relevant, as it deals with location-scale regression type modeling. See McLachlan and Basford (1988) for an attempt to adapt these methods to the multivariate normal mixture problem.

There seem to be two other feasible approaches. One is to exploit special structures of the densities involved to create diagnostic procedures and goodness-of-fit tests, thereby constructing, in stages, a model suitable to the data at hand. Another approach is to use, in conjunction with maximum likelihood, a more robust procedure based on minimum distance ideas. Both areas have seen relatively little development in the mixture model.

1.4.2. *Extracting signal from noise.* Another important warning relates to understanding that we cannot possibly discern very much of the fine detail about the distribution Q. In particular, estimating its density $dQ(\phi)$ with a realistic sample size is virtually impossible. Moreover, it is not uncommon that the goodness-of-fit of a mixture model to a data set does not change very much if we switch from a continuous latent distribution to a discrete one, or whether the discrete distribution has two components or four components.

To make this point more clearly, we consider the very simplest of mixture scenarios. Suppose we have a mixture of two normals, say

$$g(x) = \pi N(-a, 1) + \bar{\pi} N(+a, 1).$$

Assume for the moment that a is fixed and known, so that the only unknown parameter is π. If the variance of the normals were very small, the resulting data would appear much like a Bernoulli distribution, in which nearly all the observations would be quite close to $\pm a$. If the data were exactly Bernoulli, the Fisher information about π in a single observation would be $1/\pi\bar{\pi}$. Thus we know that for reasonable accuracy in estimating π when it is near 0.5, one would need a sample of roughly Gallup poll size, say 1000, yielding a standard error of $1/\sqrt{4000} \approx 0.02$.

How much information is in the mixture distribution? As an *exercise*, check the following calculations. The information in π at $\pi = 0.5$ has the form, writing the normal density as n,

$$i_a = E\left[\frac{n(X; a, 1) - n(X; -a, 1)}{0.5n(X; a, 1) + 0.5n(X; -a, 1)}\right]^2.$$

The information relative to the ideal information $1/\pi\bar{\pi} = 4$ is then $i_a/4$. This relative information has been plotted as a function of $2a$, the separation of the means, in Figure 1.3.

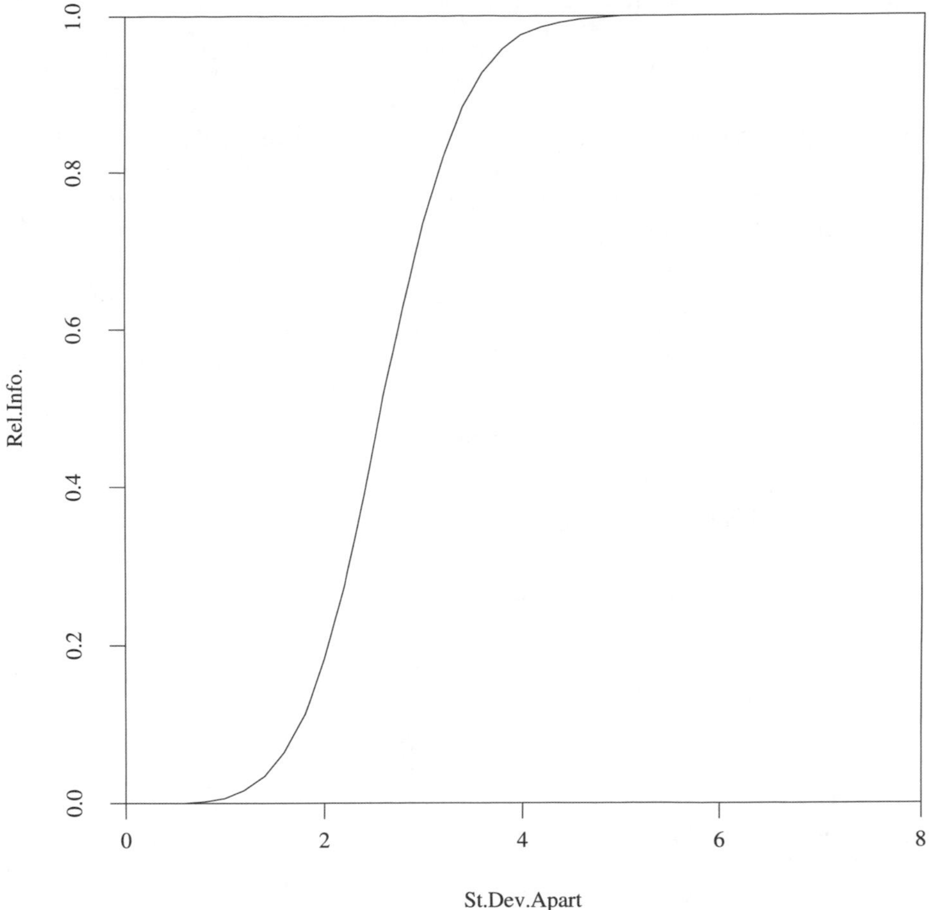

FIG. 1.3. *The information about π as a function of the separation of the means.*

We can see that the relative information is very nearly 1 when the means are four or more standard deviations apart. We can interpret this as showing that we can infer the group label J from the observation X quite accurately and that the mixture maximum likelihood estimator of π is very nearly equal to $n^{-1}\sum J_i$. However, as we move the difference in means from four to two standard deviations, the information falls off dramatically, until at two standard deviations separation we see that in order to attain the same accuracy of estimation as in the Bernoulli trials situation, it would take a sample size *10 times* as large. Recall that, as shown in Figure 1.2, this is exactly where the density is on the unimodal–bimodal boundary. When the separation of means is between 0 and 2, there is virtually no information about π.

This analysis is clearly relevant to the problem of estimating a latent density function $dQ(\phi)$, as indicated by our difficulty in separating out the relative contributions π of two nearby values of ϕ. Indeed, by setting $a = h$ and letting h go to zero, we have

$$i_h = 4h^4 E\left[\frac{n''(x;0)}{n(x;0)}\right]^2 + o(h^4).$$

The absence of terms of lower order than h^4 is another indicator of the small amount of information in this model when the components are close together. The term inside the expectation will show up again when we consider the $C(\alpha)$ test in Chapter 4.

(If we modified this problem so that the two normal location parameters were unknown, then clearly the information about π is smaller. However, the order of magnitude is not changed from the above calculation.)

This low level of information is also an indication that algorithmic methods will have difficulty. In particular, the EM algorithm will be very slow in this situation.

1.5. The likelihoods. In the examples above and in many typical examples, we will be concerned with independent samples for which the likelihoods have the form

$$L(Q) = \prod_{i=1}^{n} L_i(Q),$$

where $L_i(Q)$ has the integral form $\int L_i(\phi)\,dQ(\phi)$. The corresponding log likelihood will be denoted

$$l(Q) = \ln(L(Q)) = \sum \ln(L_i(Q)). \tag{1.4}$$

The standard analyses of a model with mixture structure involves making one of two assumptions:

- There are a fixed and known number of components, so that the unknown parameters are the π's and ϕ's.
- The distribution Q is from some known parametric family of distributions, say $q(\phi;\gamma)$, typically chosen to be either the normal density with unknown mean and variance or the conjugate family of distributions to the density f.

As a basis for statistical likelihood procedures, both standard modeling methods have their drawbacks. For the fixed component models, there is the problem that the likelihood is high dimensional and known to be multimodal. (A further awkwardness, which we will study in depth in Chapter 4, is the aberrant behavior of the limiting distribution of the likelihood ratio statistic when we test for the number of components.) In the continuous case, numerical integrations are usually needed to carry out the procedure and, regardless of this, one can end up again with a multimodal likelihood.

If, on the other hand, we make no restrictions on the number of components, but view the above likelihood as a function of a completely unknown distribution function Q, then the solution has some extremely nice properties, as will be discussed shortly. Before proceeding on this point, we introduce some special likelihood structures.

1.5.1. *The multinomial likelihood.* First, a number of simplifications in the study of mixture models occur when the component densities are discrete, so that if the sampling is i.i.d., we have a *multinomial likelihood.* Suppose that we have an i.i.d. sample from a discrete mixture density function $f(t; Q) = \Pr(X = t; Q)$ with range $t = 0, \ldots, T$. Then the likelihood of a sample $X_1, \ldots, X_n$ can be written as

$$L(Q) = \prod_{i=1}^{n} f(x_i; Q) = \prod_{t=0}^{T} f(t; Q)^{n(t)}.$$

Here $n(t)$ is the count of the number of the X's that took on value t and it is clear from the representation of the likelihood that the values $n(0), \ldots, n(T)$ are sufficient statistics for the parameter Q. The log likelihood (1.4) can therefore also be written as

$$l(Q) = \sum_{t=0}^{T} n(t) \ln(f(t; Q)). \tag{1.5}$$

1.5.2. *Partly classified data.* Another likelihood structure that occurs in many examples of interest arises when, in a sampling situation with physically identifiable components, the data are *partly classified.* That is, a portion of the sample, the *unclassified* part, has the variable J_i missing and so has the mixture model structure. In the remainder of the sample the J_i are *not* missing and so this group could be identified as being *fully classified.*

In this case the likelihood will be a product of two terms: the first, say $L_1(Q)$, being the usual mixture likelihood for the first sample, whereas the second will have the form

$$L_2(Q) = \prod P(X_i = x_i | J_i = j_i) P(J_i = j_i) = \prod f_i(x_i; \xi_j) \pi_{j_i}.$$

We next show that one can write this likelihood in the mixture form (1.4) and so this complication of the data structure does not require any additional theory.

We start by defining the likelihood kernel

$$L_i(\phi) := f_i(x_i; \phi) \mathscr{I}[\phi = \xi_{j_i}]. \tag{1.6}$$

Here the symbol $\mathscr{I}[\cdot]$ represents the indicator function, a function of all the arguments included in its brackets, which takes on the value 1 when the bracketed statement is correct and 0 when it is false. Note that although L_i is a function of the data, we can and do express this only through the subscript i. (We can do so because in a likelihood, the data are fixed.) Moreover, the corresponding term in the likelihood can be written as an integral over the mixing distribution Q: As an *exercise*, check that for Q discrete we can write

$$L_2(Q) = \prod f_i(x_i; \xi_{j_i}) \pi_{j_i} = \prod \int L_i(\phi) \, dQ(\phi).$$

For the theory of maximum likelihood we will present, it is not mandatory that L be a density function—only that it is nonnegative, a prescription satisfied by L_i in this case.

A similar likelihood situation arises when the latent variable $\Phi_i = \phi_i$, possibly continuous, has been directly measured in a subsample. In this case the appropriate likelihood kernel analogue of (1.6) is

$$L_i(\phi) := f_i(x_i;\phi)\mathscr{I}[\phi = \phi_i].$$

The distinction between (1.6) and this case is that in the former the exact value of the latent variable Φ_i is not observed; all that is known is that it is the value of the parameter ξ associated with the j_i component. Thus (1.6) has less information about the latent distribution than when the latent variable is itself observed.

1.6. The mixture NPMLE theorem. The idea of finding a nonparametric maximum likelihood estimator (NPMLE) of a latent distribution is an old one. The idea was suggested in an abstract by Robbins (1950), and later received substantial theoretical development by Kiefer and Wolfowitz (1956). Although the latter showed that the method had great theoretical promise as a method of providing consistent estimators in problems with many nuisance parameters, there was no development of numerical methods for the computation of such an estimator. The development of such methods, together with further properties of the estimator $\hat{Q}$, arose in papers that came 20 years and more later, particularly Simar (1976), Laird (1978), Jewell (1982) and Lindsay (1981, 1983a, b). In this chapter we will summarize the most important developments by informally describing the "fundamental theorem of nonparametric mixture maximum likelihood estimation." We will return to the details and proofs in Chapter 5.

1.6.1. *The fundamental theorem.* We consider the problem of maximizing the objective function

$$l(Q) = \sum_{s=1}^{D} n(s)\ln(L_s(Q))$$

over all distribution functions Q, where

$$L_s(Q) := \int L_s(\phi)\,dQ(\phi)$$

and we have written the likelihood allowing for multiple observations $n(s)$ of a single L_s. We assume that the L_s are all distinct, in that no two arise from identical likelihood kernels. We assume all $n(s) > 0$, but they need not be integers.

These assumptions give D a precise meaning as the *number of distinct summands* in the log likelihood. The only substantial further assumption is that the individual kernels of the likelihood, namely, $L_i(\phi) := f_i(x_i;\phi)$, are both nonnegative and bounded as functions of ϕ. There do exist models with unbounded likelihoods for which the theory is therefore inapplicable without some modification.

Technically, assuring existence does require another two assumptions that will be discussed in Chapter 5, but we know of no cases where they present a genuine difficulty.

PART 1. *Existence and discreteness.* The first part of the theorem states that, under the assumptions above, there *exists* a maximum likelihood estimator (MLE) $\hat{Q}$ that is a *discrete* distribution with no more than D distinct points of support ξ_r. This bound guarantees, of course, that the number of support points is no more than n, the sample size.

The implication of this part of the result is that we can carry out the calculation of this estimator using known techniques from the theory of finite mixture models and that there is an upper bound on the complexity of this distribution.

PART 2. *Gradient characterization.* The second part of the theorem gives us a way of testing whether a given latent distribution, say Q_0, is the MLE. The idea is relatively simple. Suppose we have a distribution Q_0 that is a candidate to be the MLE. We cannot determine if it is the MLE by a direct search process because the space of distributions is infinite dimensional.

However, there is a simple function we can calculate to determine if the solution has been found. We first form a *path* in the space of distribution functions from Q_0 to any other distribution, say Q_1, by letting $Q_\alpha = (1-\alpha)Q_0 + \alpha Q_1$. For every α, this generates an intermediate distribution, with $\alpha = 0$ and 1 corresponding to the original two distributions of interest.

Next, we compute the likelihood along this path, obtaining a one parameter likelihood function $L^*(\alpha) = L(Q_\alpha)$. The derivative of $\ln(L^*(\alpha))$ at $\alpha = 0$ is the *directional derivative* corresponding to this path from Q_0 to Q_1 and [*exercise*] it has the simple form

$$D_{Q_0}(Q_1) = \sum_{i=1}^{D} n(i)\left(\frac{L_i(Q_1)}{L_i(Q_0)} - 1\right).$$

A special case of this derivative occurs when we look along paths where Q_1 is degenerate; that is, a point mass at a single point ϕ, which we will denote Δ_ϕ. We define the *gradient function* to be

$$D_{Q_0}(\phi) := D_{Q_0}(\Delta_\phi) = \sum_{i=1}^{D} n(i)\left(\frac{L_i(\phi)}{L_i(Q_0)} - 1\right). \tag{1.7}$$

Note that

$$D_{Q_0}(Q_1) = \int D_{Q_0}(\phi)\, dQ_1(\phi).$$

[*Exercise.*] This has the important implication that once the gradient function has been determined, we can also determine the value of the directional derivative toward *any* distribution Q_1 by integrating the gradient function.

Next, it is clear that if the gradient function $D_Q(\phi)$ takes on positive values at any ϕ, then the likelihood along the path from Q in the direction of Δ_ϕ

is increasing at Q, so that Q cannot be the maximum likelihood estimator. However, we will show later that a much stronger result holds: that Q is a maximum likelihood estimator *if and only if*

$$D_Q(\phi) \leq 0 \qquad \forall \phi. \tag{1.8}$$

That is to say, all the information we need to find a maximum likelihood estimator is contained in the gradient function.

Moreover, if the *gradient inequality* (1.8) fails for candidate Q at some ϕ_0, not only do we learn that we are not at the maximum, but we also know one way to increase the likelihood—simply move some mass to ϕ_0. Algorithms based on this idea will be discussed in Chapter 6.

PART 3. *Support point properties.* The third part of the theorem regards the location of the support points $\hat{\xi}_j$ of $\hat{Q}$. The result is that if ξ is a support point for any $\hat{Q}$ that maximizes the likelihood, then

$$D_{\hat{Q}}(\xi) = 0.$$

Together with the gradient inequality (1.8) this implies that the support points will be local maxima of the gradient function $D_{\hat{Q}}(\phi)$. One of the consequences of this result is that a gradient-based algorithm need not keep track of the support points in $\hat{Q}$ because they can be recovered from the gradient function at the end of the algorithm. This result is also very useful in proofs of the uniqueness of the MLE $\hat{Q}$.

PART 4. *Uniqueness.* The final main result is that the *fitted values* of the likelihood, namely,

$$\mathbf{L}(\hat{Q}) := (L_1(\hat{Q}), \ldots, L_D(\hat{Q})),$$

are uniquely determined. That is, even if there were two distributions maximizing the likelihood, they would generate the same vector of likelihood fitted values. This elementary geometric result, together with Part 3, can be extended by a much more sophisticated analysis to prove the uniqueness of $\hat{Q}$ within various families of component distributions. See Chapter 5.

AN IMPORTANT REMARK. One the most striking features of the above theory is the complete lack of regularity conditions on the models and the complete generality with regard to the parameter space of ϕ.

1.7. Related nonparametric problems. We earlier introduced some classical nonparametric models with hidden mixture structure. We will use them here to illustrate the workings of the NPMLE of a latent distribution.

1.7.1. *The MLE of an unknown distribution.* We start with the simplest, but still profound, result. Suppose we have a sample $x_1, \ldots, x_n$ from a completely unknown distribution F. Suppose further that we can restrict attention to distributions that are discrete. That we can do so is not obvious, but a number of authors have developed approaches to nonparametric maximum likelihood that lead to this conclusion in this problem. See, for example, Scholz (1980).

With this step taken, the likelihood can be meaningfully written as $L(F) = \prod F(\{x_i\})$, where we use the convention of using the symbol F for both distribution function and measure. If we let $\pi_i = F(\{x_i\})$, then the problem is to maximize $\prod \pi_i$ while maintaining the obvious constraints on the π_i. This can be carried out by using a Lagrange multiplier for the inequality constraint $\sum \pi_j \leq 1$, which results in $\hat{\pi}_j = 1/n$. [*Exercise.*] The resulting distribution is called the *empirical distribution function* and is denoted $\hat{F}$.

The mixture version arises as follows. First, replace F by Q in the notation and let the likelihood kernel function be defined by

$$L_i(\phi) = \mathscr{I}[\phi = x_i].$$

If this is done, then $L_i(Q) = Q(\{x_i\})$ and the mixture likelihood corresponds exactly to the above likelihood. [*Exercise.*]

The latent distribution NPMLE theorem can then be used to prove that the empirical distribution function gives the maximum likelihood estimator. We need only to check the gradient inequality and it is easily verified [*exercise*] that

$$D_{\hat{F}}(\phi) = \begin{cases} 0, & \text{if } \phi = x_i \text{ for some } i, \\ -n, & \text{else.} \end{cases}$$

Note that, in accordance with Part 3, the estimated support points are local maxima of the gradient.

1.7.2. *Accurate and error-prone measurements.* The preceding example leads naturally to the appropriate technique to use when some of the ϕ's are seen directly, say $\phi_1, \ldots, \phi_a$, and others are observed indirectly, through $X_i | \Phi = \phi_i$. This occurs, for example, in some measurement error problems, where in order to ascertain the level of the measurement error, on a small subsample one takes much more accurate measurements (presumably also more expensive and/or time-consuming). The observed ϕ's are then assumed to be the gold standard whose relationship to response variable Y is desired. In such a case, as we have earlier indicated, we can write the likelihood in two parts, one part being from the gold standard measurements and so it has indicator functions for the likelihood kernel, as in the preceding example, and the second part has the mixture form corresponding to the density of X given Φ. See Roeder, Carroll and Lindsay (1993).

1.7.3. *Monotone density problems.* We have already indicated that the class of nonincreasing density functions on $[0,\infty)$ is a model with hidden mixture structure.

It is well known that the nonparametric MLE of such a nondecreasing density function can be characterized as having a compound distribution function (CDF) $\hat{G}$ that is the "least concave majorant" of the empirical distribution function [e.g., Groeneboom and Wellner (1992)]. That is, $\hat{G}$ is that concave function whose graph lies above that of $\hat{F}$, but is closest to it. The graph is piecewise linear; its points of contact with $\hat{F}$ are the points where it bends and these points are some subset of the observations.

It can be checked that this description implies that the nonparametric MLE of the latent distribution Q has its mass at some subset of the observations and so the mixture solution has the representation $\sum \pi_i U(0, x_i)$, where U indicates a uniform distribution on the specified range.

The gradient characterization of this problem can be used to obtain this solution. The reader may find it a helpful *exercise* to do so, following this line of argument. First, by examining the gradient it can be determined that it must have all its local maxima at the observations x_i, so, by Part 3 of the theorem, the support points must be among this set. Next, it can be shown that if ϕ_1 is the smallest support point of the latent distribution, then $f(x;\hat{Q})$ must be constant on the interval $(0,\phi_1)$, and the gradient inequality on the interval $[0,\phi_1]$ implies that $\hat{F}(t) \leq F(t;\hat{Q})$, for $t < \phi_1$, with equality at $t = \phi_1$ (since it is a support point). Since $F(t;\hat{Q})$ is linear on this interval and must be concave overall, it is clear that this defines the first support point as corresponding to that point $(x_i, \hat{F}(x_i))$ first intersected by a ray from the origin that is rotated from the y axis toward the x axis. Thus we have shown that the solution "majorizes" the empirical distribution function on this first interval. We can then continue to the next support point ϕ_2 and slightly modify this argument to show majorization over the interval $[\phi_1, \phi_2]$, with equality at the endpoints.

1.7.4. *Censoring problems.* Another important class of nonparametric problems that have hidden mixture structure arise in censoring problems. For example, consider the problem of finding the distribution function F that maximizes the likelihood $L(F) = \prod F(\{x_i\}) \prod F([c_j,\infty))$. This is the likelihood that arises under so-called noninformative right censoring, in which the x_is correspond to observed lifetimes, but all that is known about the observations in the second set is that they fell to the right of the censoring values c_j.

We can turn this into a mixture problem, as before by using indicator functions, where we now use $\mathscr{I}[\phi = x_i]$ for the observed data and $\mathscr{I}[\phi \geq c_j]$ for the censored data. Once again we can show that the gradient function has all its local maxima for ϕ in the observed data set, so the support points can be restricted to this set. The solution is the product limit estimator, also known as the Kaplan–Meier estimator. [*Exercise*: Use the gradient characterization to derive this result.]

Jewell, Malani and Vittinghoff (1994) have shown that the mixture NPMLE theorem can be used in much more complicated interval censoring problems that arises in various AIDS studies.

1.8. Similar statistical problems. As a final note to this discussion of the vast range of the mixture problem, it should also be pointed out that there are still more statistical areas that are closely related mathematically and so they carry techniques and theory that are relevant to the study of mixture models.

The mixture NPMLE theorem is, in a mathematical sense, simply a restatement, with statistical interpretation, of a basic result in the maximization of a concave objective function over a convex set. The theory of optimal design [Silvey (1980)] hinges on exactly such an optimization and it has a theorem of exactly the same form, but with other interpretations. The algorithmic literature from optimal design theory can be carried directly over to the mixture problem with slight modification.

The theory of order-restricted inference [Robertson, Wright and Dykstra (1986)] also has large areas of overlap. For example, the monotone density problem of the previous section is an example of an estimation problem carried out under a order restriction. Those restrictions often can be expressed in a way involving convexity, and the estimation problem again relates to finding the minimum or maximum of a functional over a convex set. Some of the relationships between the two will be made clearer when we deal with the likelihood ratio problem in depth in Chapter 4.

CHAPTER 2

Structural Features

This chapter is devoted to developing a mathematical understanding of the structures that are inherent to the mixture model, ranging from the simple properties of moments up to rather complicated features of exponential family mixtures. Sections 2.1 and 2.2 contain the material of greatest practical importance because they address features of the mixture model useful for diagnostic purposes. The material thereafter is very important for understanding the issues of identifiability of the latent distribution Q, but can be skimmed and returned to as needed for the later chapters.

2.1. Descriptive features.

2.1.1. *Some simple moment results.* One of the nicest mathematical features of the mixture model is the simple way in which the latent distribution Q enters into the calculation of expectation. Simply by reordering the order of integration (or summation), we obtain the fact that if $t(x)$ has expectation $\tau(\phi)$ under the unicomponent model $f(x;\phi)$, then it has expectation $\int \tau(\phi)\, dQ(\phi)$ under the mixture model $f(x;Q)$. This is easily shown using the latent variable Φ:

$$E[t(X);Q] = E[E[t(X)|\Phi]] = E[\tau(\Phi)].$$

Using the latent variable also simplifies the calculation of variances under the mixture model:

$$\text{Var}[t(X);Q] = \text{Var}(E[t(X)|\Phi]) + E(\text{Var}[t(X)|\Phi]). \tag{2.1}$$

To illustrate these formulas, suppose that X comes from a mixture of Poisson densities with mean parameter ϕ. Then the following simple relationships between the marginal mean and variance of X and the latent variable Φ hold:

$$\begin{aligned} E(X;Q) &= E[\Phi], \\ \text{Var}(X;Q) &= \text{Var}[\Phi] + E[\Phi]. \end{aligned}$$

[*Exercise.*] Manipulation of these equations then shows that the variance of X in a Poisson mixture model is inflated, compared to a unicomponent Poisson model, in the sense that the *variance-to-mean ratio* is inflated to a value larger than the ratio 1 of the unicomponent model.

Examination of the variance formula (2.1) shows that there is a general sense in which the mixture model will create extra variation over the unicomponent models that generate it. This section of this chapter is devoted to two important diagnostic features of the mixture model related to the way the extravariability will show up in the observable distribution of X:

- There is a tendency for the presence of the mixture model to be evident in the form of multimodality.
- A comparison of a multicomponent mixture distribution with the unicomponent model yields a very strong form of stochastic ordering between the two, related to the heavier tails the mixture distribution will have.

2.1.2. *Shape and modality.* We have already seen that in the normal mixture model, having two components is not synonymous with having two modes. For more details on the exact conditions under which the two-component normal mixture is bimodal, see Robertson and Fryar (1969).

Thus if we were to examine a histogram of data that were unimodal, we could not discard the possible presence of two or more normal components mixed together. However, by considering ratios of densities, we can greatly increase the sensitivity of plots. Suppose that

$$g_2(x) = \bar{\pi} n(x; \phi_1, \sigma^2) + \pi n(x; \phi_2, \sigma^2)$$

is a two-component normal mixture density, with

$$\text{mean } E[X] = E[\Phi] \text{ and variance } \text{Var}(X) = \sigma^2 + \text{Var}[\Phi].$$

Let $g_1(x) = n(x; E[X], \text{Var}(X))$ be the unicomponent normal density with the same mean and variance as the two-component mixture. Lindsay and Roeder (1992b) show that the ratio $g_2(x)/g_1(x)$ is *always* bimodal. Moreover, as x goes from $-\infty$ to $+\infty$, the centered ratio

$$\frac{g_2(x) - g_1(x)}{g_1(x)}$$

will have the sign sequence $(-, +, -, +, -)$, reflecting four crossing points of the density functions g_1 and g_2.

The remarkable feature of this result is that the bimodality and sign changes occur even if the component densities are arbitrarily close together, and no matter the magnitude of π.

We will not give a complete proof of this result here nor a detailed look at how one can use this for a diagnostic in the normal problem. For the latter, see Roeder (1994). The complete proof relies on results in totally positive kernels and some properties of moment-matched distributions. The interested reader will find that the technical material later in this chapter provides much of the background needed for understanding the proof.

2.1.3. *Overdispersion and sign changes.* It is very common to find that the component densities used in a mixture model come from an exponential family. This is fortunate because there is much that can be said about the structural features of such exponential family mixtures. We have already noted that in the normal family we can be very precise about the relationship between the densities of the two-component and one-component models that have the same mean and variance. We will now develop similar results for the *one parameter exponential family*.

We suppose that the component densities have the form

$$f(x;\phi) = \exp(\phi x - \kappa(\phi)) \tag{2.2}$$

with respect to some supporting measure $dF_0(x)$ on $\mathbf{R}^1$. The supporting measure contains all parts of the density not depending on ϕ and can be chosen to be one of the members of the family.

For the reader unfamiliar with these representations, we give an example of a statistical model that can be put into this canonical form. For example, in the binomial distribution with sample size parameter n and success parameter p, which we denote $\text{Bin}(n, p)$, we can let $dF_0(x)$ be the binomial distribution with $p = 0.5$. To write the density in the canonical form, ϕ is the log odds parameter $\ln[p/(1-p)]$ and

$$\kappa(\phi) = n\ln(1-p) - n\ln(2).$$

Shaked (1980) established a number of important properties of exponential family mixtures. For a given latent distribution Q, suppose that $f(x;\phi_0)$ is the unicomponent model that has the same mean for X as does the mixture distribution; that is, $E[X;\phi_0] = E[X;Q]$. Define the *ratio function* by

$$R(x) = \frac{f(x;Q)}{f(x;\phi_0)}. \tag{2.3}$$

The key results are:

- $R(x)$ is a convex function of x.
- $R(x) - 1$ has the sign sequence $(+, -, +)$ as x traverses the real axis.

We will prove these results in the next subsection.

These properties show that the mixture density has heavier tails than the mean-matched unicomponent model. Shaked used the convexity result to show a type of stochastic ordering between the multicomponent distribution and the corresponding unicomponent distribution:

the distribution F_Q is a dilation of F_{ϕ_0}.

Here we say that a distribution G is a *dilation* of distribution F if $\int x\,dF(x) = \int x\,dG(x)$ and if, for every convex function $c(x)$,

$$\int c(x)\,dG(x) \geq \int c(x)\,dF(x).$$

Notice that Shaked's dilation result implies directly the overdispersion result

$$\text{Var}(X; Q) \geq \text{Var}(X; \phi_0).$$

[*Exercise.*]

We note further that G is a dilation of F if and only if they have the same mean and there exists a family of distributions $K(y|x)$, with $\int y\, dK(y|x) = x$, such that $G(y) = \int K(y|x)\, dF(x)$. In our case, this means that there exists $K(y|x)$ such that

$$F_Q(y) = \int K(y|x)\, dF_{\phi_0}(x).$$

Thus in the mixture case we can think of the mixture variable Y as being generated in two steps: first generate $X = x$ from the corresponding unicomponent model, then dilate it by generating Y from a kernel distribution with mean x. This, curiously, reverses the original representation, putting the unicomponent distribution in the role of the latent distribution and K in the role of the component density family.

As an illustration of how these dispersion properties show up in data, we consider the data set in Table 2.1, which identifies the number of male children in 6115 sibships of size 12, collected in Saxony, Germany [Geissler (1889)].

Given a parenting couple, we might model their children as being born with independent sex determination, like a sequence of Bernoulli trials with some constant probability p of having a male child. If so, the number of male children X in a family with 12 children would be distributed as a Bin$(12, p)$ random variable. In this context it is natural to ask if the probability of a male birth p is a latent variable, varying from family to family, or is constant across families. If p does vary, then we could associate with each couple a

TABLE 2.1

Number of male children in sibships of size 12

# Males	Obs. Count	Obs. vs. Fit	Bin. Fit
0	3	>	0.9
1	24	>	12.1
2	104	>	71.8
3	286	>	258.5
4	670	>	628.1
5	1033	<	1085.2
6	1343	<	1367.3
7	1112	<	1265.6
8	829	<	854.3
9	478	>	410.0
10	181	>	132.8
11	45	>	26.1
12	7	>	2.3
Total	6115		6115.0

latent parameter p_i representing their propensity to have male children. The result is that the data would be from a mixture of binomials.

Table 2.1 also shows the expected values of those counts, assuming that they arose from sampling from a unicomponent binomial distribution $\text{Bin}(12, p)$, where p was estimated from the sample to be 0.51. It is clear that the observed distribution has heavier tails than would be expected from the unicomponent binomial model and, moreover, that the difference *observed* − *expected* has the sign change behavior $(+, -, +)$ predicted under the above results for a mixture model.

2.1.4. *Log convexity of ratios.* Establishing the first part of Shaked's results regarding the convexity of $R(x)$ is simple and instructive. Indeed, we can just as easily show a stronger result, that

$$\ln(R(x)) \text{ is a convex function of } x. \tag{2.4}$$

(Why is this stronger?) We learn this by examining the structure more closely. We can write

$$\begin{aligned} R(x) &= \frac{\int \exp(\phi x - \kappa(\phi))\, dQ(\phi)}{\exp(\phi_0 x - \kappa(\phi))} \\ &= \int \exp((\Phi - \phi_0)x)\, d\Gamma(\Phi). \end{aligned} \tag{2.5}$$

Here the positive measure Γ is defined by

$$d\Gamma(\phi) = \exp(\kappa(\phi) - \kappa(\phi_0)) \cdot dQ(\phi).$$

This demonstrates that $R(x)$ has the mathematical structure of a *Laplace transform*, that is, it is a scalar multiple of a moment generating function. Since this means its logarithm is a scalar translation of a cumulant generating function, this implies that $\ln(R(x))$ is a convex function of x.

(*Technical note*: In an exponential structure model, such as the binomial, it is possible that there exist distributions that correspond to infinite values of the natural parameter ϕ. Thus, when the binomial success parameter p is 0, the natural parameter ϕ is $-\infty$. Including these points in the analysis would involve technical difficulties that we would rather avoid here, so we will assume Q assumes no mass at infinity, although it is not usually necessary to do so.)

Indeed, we gain further insight by differentiating twice, to find that

$$\begin{aligned} \frac{d}{dt} \ln(R(t)) &= E[(\Phi - \phi_0)|X = t], \\ \frac{d^2}{dt^2} \ln(R(t)) &= \text{Var}(\Phi|X = t). \end{aligned}$$

[*Exercise.*] That is, differentiation of the log ratio function generates the cumulants of the posterior distribution of the latent variable, given the observation.

Now the convexity of $\ln(R(x))$ implies the convexity of the centered ratio function

$$R(x) - 1 = \frac{f(x;Q) - f(x;\phi_0)}{f(x;\phi_0)},$$

so we have only to prove the sign change result, which can be done by showing that the difference $f(x;Q) - f(x;\phi_0)$ has the sign pattern $(+,-,+)$. That is, the convexity of the function R shows that there can be at most two crossings of zero, so what remains is to show that there are exactly two. Although this can be done directly, we instead take a diversion into much stronger results relating moments and sign changes. The reader may wish to skim the following section on first passage.

2.1.5. *Moments and sign changes.* The following striking result relates moments and the sign crossing properties of density functions.

PROPOSITION 1. *Suppose f and g are two density functions on $\mathbf{R}$ with supporting measure $d\mu(x)$ and possessing the same first M moments $E(X^k)$, $k = 1, \ldots, M$. Then the difference between the densities,*

$$\Delta(x) := f(x) - g(x),$$

has at least $M+1$ sign changes, unless the distributions are identical.

PROOF. Suppose not and that there are K, with $K \leq M$, nodes $t_1, \ldots, t_K$ such that $\Delta(x)$ has a constant sign between nodes, with signs alternating on adjacent internodal intervals. We can construct a polynomial

$$p(x) = \pm(x - t_1)\cdots(x - t_K)$$

of degree K that has the same sign between nodes as does $\Delta(x)$. Hence $p(x) \cdot \Delta(x) \geq 0$. However, if we integrate this function, the equality of the first K moments implies we get zero. The conclusion is that $p(x)\Delta(x) = 0$, almost everywhere $d\mu$. This shows that $\Delta(x)$ is zero (μ a.e.) between the nodes. This indicates that f and g can yield different probability measures only if they have a discrete component that differs only on the nodes. However, we can then apply the following lemma, which shows that two discrete densities having a common support set of K points, with their first K moments matching, must be identical. □

In addition to its use in the above proof, the following result is very important in the theory of the method of moments in mixture distributions [Lindsay (1989a)]. It will be useful in this chapter to have a special notation for the construction of a vector consisting of the powers of a basic variable, so we define

$$\mathbf{x}^{\dagger} = \begin{pmatrix} 1 \\ x \\ \vdots \\ x^K \end{pmatrix}.$$

LEMMA 2. *Suppose that F and G are two distributions with support on some fixed set of $K+1$ (or fewer) points $\{t_0, t_1, \ldots, t_K\}$. Further, suppose that they match in their first K moments,*

$$\int \mathbf{x}^\dagger \, dF(x) = \int \mathbf{x}^\dagger \, dG(x).$$

Then $F = G$ and the masses at the support points are given by the matrix equation (2.6) *below.*

PROOF. For any given set of moments $\int \mathbf{x}^\dagger \, dF(x)$ on a known support set $\{t_0, \ldots, t_K\}$, the corresponding masses $\boldsymbol{\pi}$ are determined by the matrix equation

$$\int \begin{pmatrix} 1 \\ x \\ \vdots \\ x^K \end{pmatrix} dF(x) = \begin{bmatrix} 1 & 1 & \cdots & 1 \\ t_0 & t_1 & \cdots & t_K \\ \vdots & \vdots & \vdots & \vdots \\ t_0^K & t_1^K & \cdots & t_K^K \end{bmatrix} \begin{pmatrix} \pi_0 \\ \pi_1 \\ \vdots \\ \pi_K \end{pmatrix}. \tag{2.6}$$

Thus if the square matrix in the equation is invertible, we have a unique solution. However, this is the well known *Vandermonde matrix*, with determinant known to be $\prod_{i>j}(t_i - t_j)$. □

Finally, we can turn the result concerning density functions into a result regarding distribution functions by using integration by parts.

PROPOSITION 3. *If two distribution functions F and G have their first K moments in common, then either they are equal or the difference $F - G$ has at least K sign changes.*

The proof is an *exercise*. See Lindsay and Roeder (1992b) for more details.

If we now return to the problem of the sign change behavior of $R(x)$, we see that since we have two densities with the same first moment, they must cross two times and the sign pattern must be $(+, -, +)$. We can additionally say that the distribution functions cross exactly once, with sign sequence $(+, -)$.

2.1.6. *Dispersion models.* Lindsay (1986) investigated in some detail the construction of parametric mixture models that had exponential family structure. We briefly survey these and related results.

One of the important uses of a mixture model is as a means of allowing for *overdispersion*. As a strategy for dealing with overdispersion, one might consider constructing an exponential family model that contains the model of interest, but contains an additional parameter to account for overdispersion. As such, consider the two parameter exponential families with densities, with respect to $dF_0(x)$, of the form

$$f(x; \alpha, \beta) = \exp(\alpha x + \beta t(x) - K(\alpha, \beta)),$$

for some function $t(x)$. Note that $\beta = 0$ generates the original model. We then ask, which functions $t(x)$ will cause this model to be overdispersed relative to the original model?

The preceding results indicate that if $f(x;\alpha,\beta)$ is to be a mixture model in the sense that there is some Q depending on α and β with $f(x;Q) = f(x;\alpha,\beta)$, then the log ratio function must be convex, from which it follows that $t(x)$ must be convex. [Lindsay (1986) gives necessary and sufficient conditions on $t(x)$.]

Gelfand and Dalal (1990) took this idea one step further and showed that if $t(x)$ is convex, even if the resulting density is not a mixture density, it still is an overdispersed density relative to the unicomponent model, in the strong sense of dilation. One example of an overdispersion model that is not necessarily a mixture model is Efron's double exponential family [Efron (1986)].

However, the more usual method for the construction of a two parameter overdispersed family is to use the *conjugate distribution*, a subject we will introduce in the next chapter.

2.2. Diagnostics for exponential families. We now turn to using the above insights for diagnostic purposes. We ask the question, for a given set of data, does the mixture model fit and, if so, do we need to use more than one component? We use the information that the ratio function is convex and also log convex. For more details on this section, see Lindsay and Roeder (1992a).

2.2.1. *Empirical ratio plots.* In a discrete sample space, with observed proportions $\hat{p}(t)$, it is natural to attempt to estimate the ratio function $R(t)$ [see (2.3)] by its empirical counterpart:

$$\hat{R}(t) = \frac{\hat{p}(t)}{f(t;\hat{\phi})}.$$

Here $\hat{\phi}$ is the maximum likelihood estimator of ϕ in the unicomponent model. Such an estimation clearly relies on the sample size being sufficiently large that $\hat{p}(t)$ is a good estimator of the true density. If the unicomponent model is correct, then the empirical ratio converges to 1 for every t. If the alternative of a mixture model is correct, then the empirical ratio will converge to the *convex* ratio function $R(t)$.

This suggests plotting $(t, \hat{R}(t) - 1)$ or $(t, \ln(\hat{R}(t)))$ and examining the plot for convexity. If the plot is clearly nonconvex, then the mixture model cannot possibly fit well. If the plot is nearly linear, then the unicomponent model is likely to fit well and strict convexity is diagnostic for mixture structure. We note that convexity is a very particular prediction for the shape of the plot, and it is something easy to identify visually. Such an empirical ratio plot is given for the data of Table 2.1 in Figure 2.1.

2.2.2. *Gradient function plots.* A second diagnostic method for looking for mixture structure is to consider the gradient function $D_Q(\phi)$ introduced in Chapter 1. One advantage to this approach is that it enables us to create plots similar to the empirical ratio plot even when the data are not discrete.

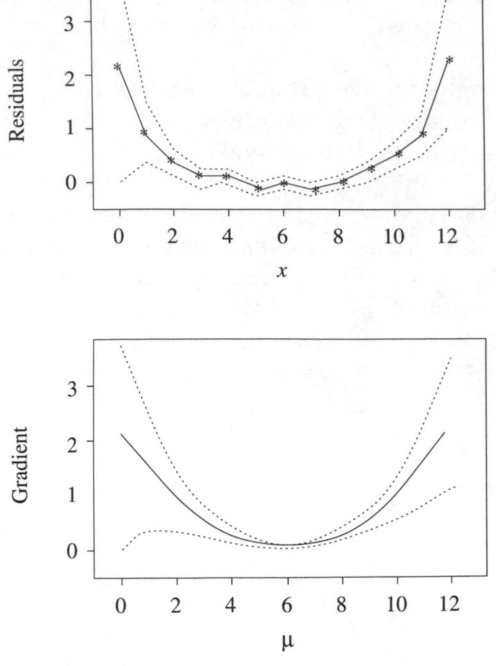

FIG. 2.1. *The ratio and gradient plots for the sibship data.*

We evaluate the gradient at $Q = \Delta_{\hat{\phi}}$, the best fitting one-component model, calling

$$D_1(\phi) := D_{\Delta_{\hat{\phi}}}(\phi)$$

the *unicomponent gradient function*. We know from the NPMLE theorem that this gradient function is diagnostic for whether or not the degenerate distribution $\Delta_{\hat{\phi}}$ is the maximum likelihood estimator of the latent distribution $\hat{Q}$. In fact, if

$$D_1(\phi) \leq 0 \qquad \forall \phi,$$

then the one-component model fits better, in the sense of higher likelihood, than any mixture model with *any* other number of components. On the other hand, if the gradient inequality is violated, we know we can increase the likelihood by adding components. This suggests that examination of $D_1(\phi)$ may be a useful diagnostic for the presence of a mixture alternative to a unicomponent model.

It is somewhat surprising that we can be precise about the predicted shape of the graph of the unicomponent gradient under the mixture model. The reason for this is most clear in the discrete case, where we can draw a close relationship between the unicomponent gradient D_1 and the empirical ratio function $\hat{R}(t)$:

$$\begin{aligned} D_1(\phi) &= \sum_t n(t)\left(\frac{f(t;\phi)}{f(t;\hat{\phi})} - 1\right) \\ &= n\sum f(t;\phi)\left(\frac{\hat{p}(t)}{f(t;\hat{\phi})} - 1\right) \qquad (2.7) \\ &= n\sum f(t;\phi)\cdot[\hat{R}(t) - 1]. \end{aligned}$$

That is, the unicomponent gradient function can be represented as a kernel-smoothed version of $[\hat{R}(t) - 1]$, where the smoothing kernel is $f(t;\phi)$. This suggests that since the empirical ratio function is asymptotically convex under the mixture model, then perhaps the gradient function could be as well.

To make this statement completely correct, we have to find the correct reparameterization of ϕ in which to plot the gradient. In order to make the result more general, we drop the multinomial assumption and consider the asymptotic limit of the unicomponent gradient function when the mixture model is correct. Check that if Q is the latent distribution and $\hat{\phi} \to$ some ϕ_0, depending on Q, that

$$\begin{aligned} n^{-1}D_1(\phi) \to D^*(\phi) &:= \int \frac{f(x;\phi)}{f(x;\phi_0)} f(x;Q)\,dF_0(x) - 1 \\ &= \int R(x)f(x;\phi)\,dF_0(x) - 1. \end{aligned}$$

Recalling that the ratio function $R(x)$ is a convex function of x, we then ask: When will a convex function $R(x)$, smoothed by a kernel $f(x;\phi)$, yield a convex function $D^*(\phi)$? When $f(x;\phi)$ is an exponential family, it will happen when the parameter ϕ is the mean value parameter of the exponential family. That is, if we replace ϕ with $\mu(\phi) = E[X;\phi]$, then

a plot of $(\mu, D^*(\mu))$ is convex.

The proof of this is relatively simple, but requires high powered results from total positivity: see Lindsay and Roeder (1992a) for details, including the calculation of statistical error bounds for the plots.

The gradient plot for the sibship data can be found in Figure 2.1, together with error bounds (calculated pointwise). Notice that it appears very much to be a smoothed version of the residual plot.

Further justification for examining the residual plot can be given by considering the *normalized gradient function*, where we divide the gradient function by its asymptotic standard error under the unicomponent model. In Chapter 4 we will show that the likelihood ratio test statistic for one component versus two components is asymptotically equivalent to the square of the maximum of the normalized gradient function, so that, in terms of the gradient plot, the likelihood ratio test is equivalent to rejecting the unicomponent model if the gradient crosses the upper confidence line. Methods for adjusting the critical value for simultaneous inference are given in Chapter 4.

TABLE 2.2
Number of male children in sibships of size 8

# Males	**Obs. Count**	**Obs. vs. Fit**	**Bin. Fit**
0	215	>	165.22
1	1,485	>	1,401.69
2	5,331	>	5,202.65
3	10,649	<	11,034.65
4	14,959	>	14,627.60
5	11,929	<	12,409.87
6	6,678	>	6,580.24
7	2,092	>	1,993.78
8	342	>	264.30
Total	$n = 53,680$		53,680.00

2.2.3. *Comparing gradient and ratio plots.* The residual plot requires a larger data set because it has no smoothing feature, but as a consequence it reveals more structure. One striking example of this occurs when we turn to the sibship data for families of size 8, presented in Table 2.2.

This table is based on a much larger sample than Table 2.1, and it shows a striking lack of fit of the mixture model in that there is too much deviation from convexity, relative to the standard error bounds, as can be seen from the ratio residual plot in Figure 2.2.

These data were examined by Fisher (1925), who said:

> The observed series differs from expectation markedly in two respects: one is the excess of unequally divided families; the other is the irregularity of the central values, showing an apparent bias in favor of even values. No biological reason is suggested for the latter discrepancy, which therefore detracts from the value of the data.

We note that the gradient plot (Figure 2.2) captures the overdispersion, but smooths out the fine structure. Such a large data set with interesting fine detail is probably fairly unusual, so it might be anticipated that the gradient, with its smoothness, is more generally the appropriate tool. In addition, it is a natural by-product of the nonparametric approach.

Extensions of these diagnostic ideas into the domain of generalized linear models has been carried out by Lambert and Roeder (1995).

2.3. Geometry of multinomial mixtures. We have already introduced, in Chapter 1, an important subclass of the mixture model where the component densities are treated as known. We now start giving a more detailed picture of the mixture model by examining the geometric structure in these, the simplest of mixture models. This will lead to a better understanding of several statistical issues, such as the identifiability of the latent distribution and how mixture models with different numbers of components are related.

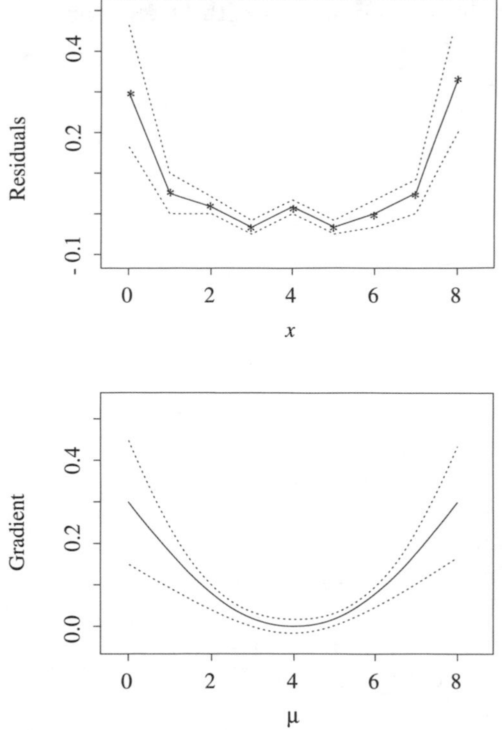

FIG. 2.2. *The ratio and gradient plots for the second sibship data.*

2.3.1. *Known component densities.* The setting is the multinomial density, with sample space $\{0, 1, \ldots, T\}$. We start with a fixed set of known component densities, say $f_1(t), \ldots, f_m(t)$, and consider mixtures of the form $f(t; Q) = \sum \pi_j f_j(t)$. We start by turning each component density function into a vector, by setting

$$\mathbf{f}_j := \begin{pmatrix} f_j(0) \\ \vdots \\ f_j(T) \end{pmatrix},$$

calling this the *density vector* for component j. Note that the entries of the vector are nonnegative and sum to 1.

Next, we need a definition. Let $\mathbf{v}_1, \ldots, \mathbf{v}_m$ be vectors in $(T+1)$-dimensional Euclidean space $\mathbf{R}^{T+1}$. If $\pi_1, \ldots, \pi_m$ is a set of nonnegative weights summing to 1, then the linear combination

$$\pi_1 \mathbf{v}_1 + \cdots + \pi_m \mathbf{v}_m$$

is called a *convex combination* of $\mathbf{v}_1, \ldots, \mathbf{v}_m$.

The fundamental result that gives power to a geometric analysis of the mixture model is extremely simple: *mixture density vectors are convex combinations of the component density vectors.*

For example, for three components with latent masses π_1, π_2, π_3, the mixture density

$$f(t; Q) = \pi_1 f_1(t) + \pi_2 f_2(t) + \pi_3 f_3(t)$$

has the vector representation

$$\mathbf{f}_Q = \pi_1 \mathbf{f}_1 + \pi_2 \mathbf{f}_2 + \pi_3 \mathbf{f}_3.$$

Moreover, if we define the matrix $\mathbf{F} = [\mathbf{f}_1, \mathbf{f}_2, \mathbf{f}_3]$, then the above relationship can be expressed as a matrix equation:

$$\mathbf{f}_Q = \mathbf{F}\boldsymbol{\pi}.$$

Thus we have a linear model, with constraints, in the parameters π_j for the mixture multinomial probabilities $\mathbf{f}_Q$.

In such a model, with known components, the mixture is *identifiable* if we can determine the values of π given the values of $\mathbf{F}$ and $\mathbf{f}_Q$. Although such a question can be addressed directly through the theory of matrices, we believe it is more insightful to use convex geometry.

2.3.2. *Basic convex geometry.* A *convex set* $\mathbf{C}$ is a set of vectors that contain every finite convex combination of its elements. Pictorially, a convex set contains all the lines connecting any two points of the set. Given a set of vectors $\mathbf{V}$, the *convex hull* of $\mathbf{V}$, denoted conv($\mathbf{V}$), is the smallest convex set containing $\mathbf{V}$.

Of particular interest to us is the case when the vectors are multinomial density vectors. That is, the vectors $\mathbf{v}$ have nonnegative entries, with entries summing to 1. Such vectors live in the *probability simplex*

$$\mathbf{P}^T = \{\mathbf{p}\colon \mathbf{p}'\mathbf{1} = 1,\ \mathbf{p} \geq \mathbf{0}\},$$

a convex set with dimension T in $\mathbf{R}^{T+1}$. (The set's dimension is reduced by 1 due to the linear constraint on the coordinates, $\sum p_j = 1$.) The probability simplex can be represented as the convex hull of $(\mathbf{e}_0, \ldots, \mathbf{e}_T)$, where $\mathbf{e}_j$ is defined to be that $(T+1)$-vector with a one in position j and zeros elsewhere. That is, the $\mathbf{e}_j$ are the usual basis vectors for Euclidean space. [*Exercise*: Sketch $\mathbf{P}^1$ and $\mathbf{P}^2$.]

We examine $\mathbf{P}^2$, a two-dimensional surface in $\mathbf{R}^3$. In Figure 2.3 we show this surface, rotated about so that it lies in the plane of the page. In addition, we show the location of the basis vectors $\mathbf{e}_j$ and a set of three multinomial component density vectors $\mathbf{f}_1, \mathbf{f}_2, \mathbf{f}_3$. From the above discussion, it is clear that the set of mixture density vectors

$$\mathscr{M} := \{\mathbf{f}_Q\}$$

is the convex hull of the set $\{\mathbf{f}_1, \mathbf{f}_2, \mathbf{f}_3\}$. The hatched set in the figure represents the possible mixture density vectors allowed under this model.

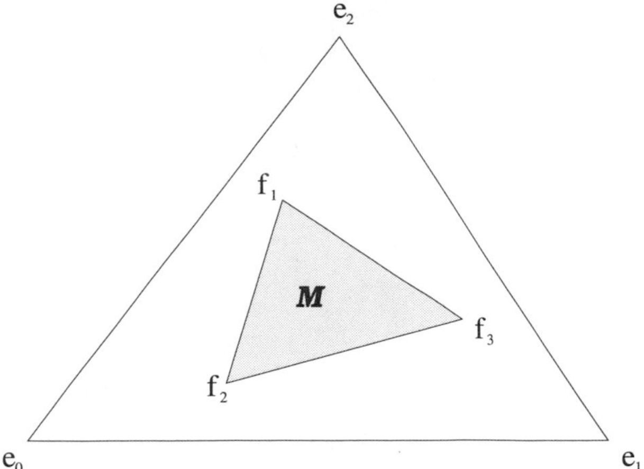

FIG. 2.3. *The two-dimensional probability simplex.*

2.3.3. *Identifiability of weight parameters.* We return to the question of identifiability. In essence, the weight parameters are identifiable if we can solve uniquely for them from the distribution of the observables, in our case from the density $f(t;Q)$ or the vector $\mathbf{f}_Q$. In a general linear problem, we can solve uniquely for the parameters $\boldsymbol{\pi}$ in $\sum \pi_j \mathbf{f}_j = \mathbf{f}_Q$ if and only if the vectors $\mathbf{f}_1, \ldots, \mathbf{f}_m$ are linearly independent. If we add the mixture model requirement that $\sum \pi_j = 1$, then the uniqueness of the solution is guaranteed under the weaker condition of *affine independence*. However, this concept need not concern us here, because when all the vectors $\mathbf{f}_j$ involved are in the probability simplex, the affine independence of the vectors is equivalent to their linear independence.

The reader should consider the geometric consequences of this. For example, in Figure 2.3 the vectors are linearly independent. A set of three density vectors that were not linearly independent would lie on a line, as in Figure 2.4. (Remember that one dimension is missing from the plot and the origin is not pictured.)

One simple conclusion from such an identifiability analysis is that the weight parameters π cannot be identifiable for all mixtures $\mathbf{f}_Q$ if the number of components m is greater than the number of multinomial categories $T+1$. That is,

$$m > T+1 \implies \text{weights not identifiable.} \tag{2.8}$$

In Figure 2.5 the reader should visualize why mixtures of four density vectors give nonidentifiable weights.

We note, however, that even if the weights π cannot all be identified, there may be identifiable linear combinations of scientific interest. See Roeder, Devlin and Lindsay (1989). Also, as we shall see, it is possible to have unique estimates of these parameters even when identifiability fails.

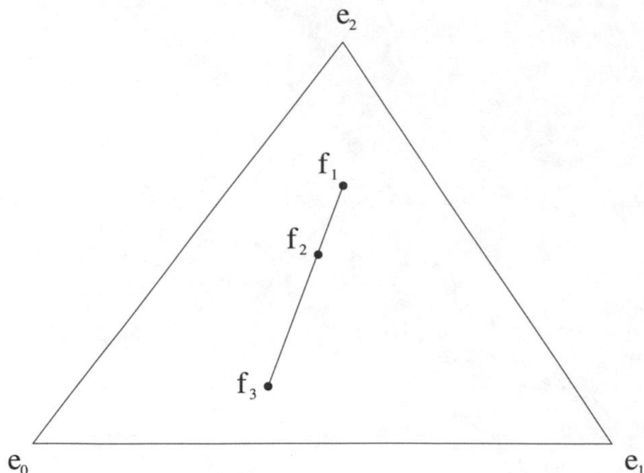

FIG. 2.4. *A mixture set with nonidentifiable weights.*

2.3.4. *Carathéodory's theorem.* One result from convex geometry that will be extremely useful to us is a classic theorem about the representation of elements of the convex hull of $\mathbf{V} \subset \mathbf{R}^K$ in terms of convex combinations of the elements of the generating set $\mathbf{V}$. *Carathéodory's theorem* says that if $\mathbf{u} \in \text{conv}(\mathbf{V})$, then there exists at least one representation of $\mathbf{u}$ as a convex combination of $K+1$ or fewer elements of $\mathbf{V}$, say

$$\mathbf{u} = \pi_1 \mathbf{v}_1 + \cdots + \pi_{K+1} \mathbf{v}_{K+1},$$

for some $\mathbf{v}_1, \ldots, \mathbf{v}_{K+1}$ in $\mathbf{V}$.

Since the density vectors involved lie in a T-dimensional subspace, it follows that there exist representations of a multinomial mixture model vector $\mathbf{f}_Q$ in terms of the convex combination of some set of $T+1$ or fewer components $\mathbf{f}_i$,

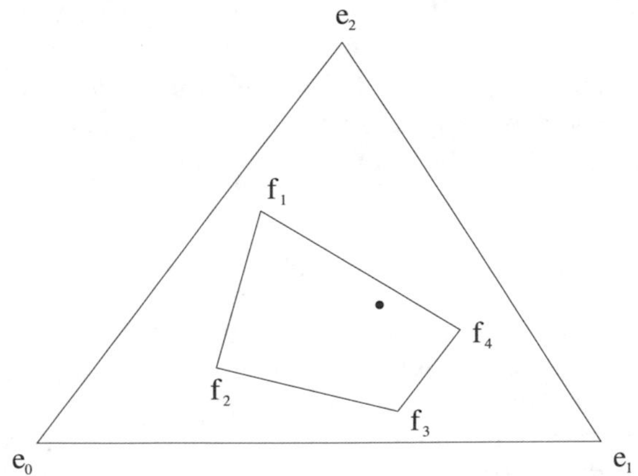

FIG. 2.5. *A second mixture set with nonidentifiable weights.*

whether or not the mixture is identifiable. As an exercise in visualization, the reader should consider the geometric truth of this theorem for models in $\mathbf{P}^2$, where it says that every point in a convex hull can be represented by three or fewer elements of its generating set.

2.4. Exponential family geometry. We now consider the question of identifiability in the one parameter exponential family model (2.2). For a multinomial exponential family, the range of the random variable X will typically be a subset of the integers of the form $\{0, 1, 2, \ldots, T\}$, with T possibly infinite. If this occurs, we will say we are in the *lattice case* and we will have special results when such structures exist.

The previous section gave us some preliminary insights into this problem, and we already know that if the range of X is finite, then there is no hope of solving uniquely for Q if it has too many support points. However, there is certainly information about the latent distribution, and the goal of this section is to identify just what structures are estimable.

2.4.1. *Identifiable functions.* A functional $h(Q)$ will be called *identifiable in the nonparametric sense* if whenever Q_1 and Q_2 are latent distributions that generate the same mixture distribution for X, then $h(Q_1) = h(Q_2)$. Thus the value of $h(Q)$ can be determined uniquely from the observable distribution of X.

In the case when X has a lattice distribution, we return to the representation of the ratio function as a moment generating function (2.5). We can conclude that, for $t = 0, \ldots, T$, the ratio function $R(t)$ is the tth moment of $\exp(\Phi)$ under the measure Γ. It follows that the first T moments of $\exp(\Phi)$ under Γ can be determined from the distribution of the observable X and so are identifiable functions of the latent distribution Q. Moreover, they are a full set in the sense that any other identifiable functional must be a function of them. [*Exercise.*]

This result is perhaps not satisfying in itself because the identifiable functions do not have a natural statistical interpretation. In some special cases, the identifiable functions have a direct interpretation in terms of the measure Q as well. For example, we have the following proposition:

PROPOSITION 4. *If X is a mixture of* $\mathrm{Bin}(T, \phi)$ *distributions, with latent distribution Q on ϕ, then the first T moments of Φ under the distribution Q are identifiable from the mixed density and all other identifiable functions are functions of these moments.*

PROOF. We prove this for the case $T = 2$; the extension to arbitrary T is an exercise. We start with the matrix identity

$$\begin{pmatrix} (1-\phi)^2 \\ 2\phi(1-\phi) \\ \phi^2 \end{pmatrix} = \begin{bmatrix} 1 & -2 & 1 \\ 0 & 2 & -2 \\ 0 & 0 & 1 \end{bmatrix} \begin{pmatrix} 1 \\ \phi \\ \phi^2 \end{pmatrix}.$$

We write this in the symbolic form $\mathbf{f}_\phi = \mathbf{A}\boldsymbol{\phi}^\dagger$, where $\mathbf{f}_\phi$ is the density vector for a binomial model with parameter ϕ and $\boldsymbol{\phi}^\dagger$ is the power vector introduced earlier. If we integrate both sides of this equation with respect to $dQ(\phi)$, we find that the left side becomes the mixture density vector $\mathbf{f}_Q$ and the right side is a linear function of the moments of Φ, which we write as $\mathbf{A}\cdot E(\boldsymbol{\Phi}^\dagger)$. Since the matrix $\mathbf{A}$ is clearly invertible, we can solve for the moments of Q, given the mixed density, via the equation $E(\boldsymbol{\Phi}^\dagger) = \mathbf{A}^{-1}\mathbf{f}_Q$. □

See Lindsay, Clogg and Grego (1991) for another example, the Rasch model, where there is a natural set of identifiable mixture parameters in the form of posterior expectations $E[g_i(\Phi)|X = x]$ that therefore allow fully identified nonparametric empirical Bayes estimation.

However, one cannot consistently estimate the latent distribution function $Q(t)$ at any value of t without further external information, such as knowing that the distribution function lies in some parametric class. In particular, the nonparametric approach of Chapter 1 cannot consistently estimate the distribution function.

Just the same, features of Q that are not identifiable are usually *partially identified*. For example, in the binomial model, knowing the first T moments of Q does limit the set of allowable distributions, the more so the larger T is. In fact, Tchebysheff developed an optimal system of upper and lower bounds for the distribution function evaluated at a point, given a set of its moments [Uspensky (1937)]. For any given functional being estimated, it is at least theoretically possible to construct upper and lower bounds that would give the degree of determination of that function.

This point is relevant statistically because it is therefore possible to construct informative confidence intervals for nonidentifiable functions of Q. Although such bounds will not shrink to zero in width asymptotically, but rather to the limits of knowledge of that function, for any fixed sample size the width due to randomness could greatly exceed that due to indeterminacy. Lindsay, Clogg and Grego (1991) consider nonparametric bounds in the Rasch model for some nonidentifiable empirical Bayes functionals of interest.

2.4.2. *Identifiability of weights, m fixed.* We leave the nonparametric setting and consider a situation where the number of components is assumed to be known, say m. The relevant question here is: If we restrict attention to latent distributions Q that have m or fewer points, will the latent distribution be identifiable, in that there will be exactly one possible latent distribution in this class that generates any one X distribution? Again, we restrict attention to discrete one parameter exponential families.

This subsection deals with the simplest case in which the support points $\xi_1, \dots, \xi_m$ of Q are known and fixed, so that our concern is with the identifiability of the weight parameters. We have already developed the appropriate basic theory for this case because this is exactly the situation of Section 2.3.1. We consider first the binomial model, for which we have the following proposition:

PROPOSITION 5. *The parameters* $\pi_1, \ldots, \pi_m$ *in the mixture*

$$\pi_1 \operatorname{Bin}(T, \theta_1) + \cdots + \pi_m \operatorname{Bin}(T, \theta_m)$$

are identifiable provided that $m \leq T + 1$. [*Note from* (2.8) *that this is the maximal number of identifiable components for this binomial family.*]

PROOF. We need to establish the linear independence of the vectors $\mathbf{f}_j := \mathbf{f}_{\theta_j}$. It therefore suffices to consider $m = T + 1$. In the notation of the proof of the last proposition, we have for an appropriate nonsingular matrix $\mathbf{A}$,

$$\begin{aligned}\det[\mathbf{f}_1, \ldots, \mathbf{f}_{T+1}] &= \det\{\mathbf{A}[\boldsymbol{\theta}_1^\dagger, \ldots, \boldsymbol{\theta}_{T+1}^\dagger]\} \\ &= \det \mathbf{A} \cdot \det[\boldsymbol{\theta}_1^\dagger, \ldots, \boldsymbol{\theta}_{T+1}^\dagger].\end{aligned}$$

It follows that it suffices to show that $\det[\boldsymbol{\theta}_1^\dagger, \ldots, \boldsymbol{\theta}_{T+1}^\dagger] \neq 0$. However, this is again the well known Vandermonde determinant, equaling $\prod_{i>j}(\theta_i - \theta_j)$. □

We next consider how this result might be extended to other one parameter exponential families. The key to the identifiability in the binomial family is the nonsingularity of the matrix of probability vectors $[\mathbf{f}_1, \ldots, \mathbf{f}_{T+1}]$, which we might believe quite difficult to deal with for an arbitrary exponential family. However, there is a quite amazing theory that relates the nonsingularity of such matrices to the maximal number of zeros of certain polynomial equations. The interested reader should dig into the difficult but impressive works of Karlin and Studden (1966) and Karlin (1968). We give a brief outline of the fundamental ideas here.

A system of functions $\tau_0(\phi), \ldots, \tau_T(\phi)$ of the real variable ϕ is called a *Chebyshev system* if every polynomial $\sum_{j=0}^{T} w_j \tau_j(\phi)$, whose coefficients w_j are not all zero, has at most T zeros in ϕ. The most familiar example of such a system is the $1, \phi, \ldots, \phi^T$, where we can apply the fundamental theorem of algebra to bound the number of real zeros by T.

Suppose we have a family of multinomial densities $f(t; \phi)$ and we define $\tau_j(\phi) := f(j; \phi)$. Further, suppose these τ_j constitute a Chebyshev system. We may conclude that

$$m \leq T + 1 \implies \text{weights identifiable},$$

any fixed set of m component densities, using the following argument:

Suppose not. Then there exists a vector $\mathbf{w}$ such that

$$[\mathbf{f}_{\phi_1}, \ldots, \mathbf{f}_{\phi_{T+1}}]\mathbf{w} = \mathbf{0}.$$

However, these equations can be written out row-by-row to show that the ϕ_j are $T+1$ solutions to $\sum w_j \tau_j(\phi) = 0$, a contradiction to the Chebyshev system property.

Although it is not insightful to our present task to prove this result, it is known that every finite discrete exponential family generates a Chebyshev system. There are other useful models that form Chebyshev systems. One

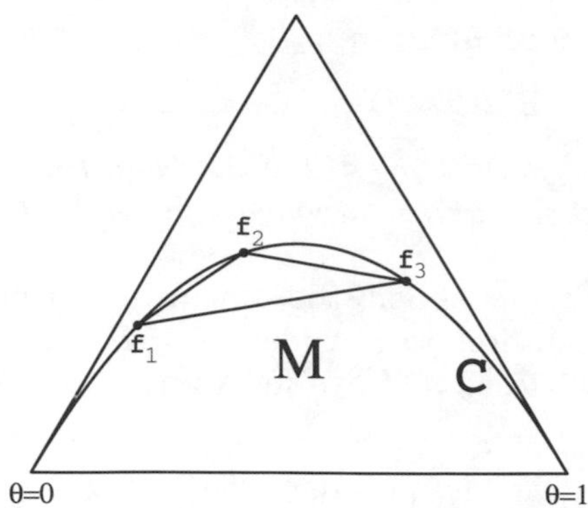

FIG. 2.6. *The binomial unicomponent density curve.*

such extension is a continuous exponential family that has been discretized into bins, as done when performing a chi-squared goodness-of-fit test. See Lindsay and Roeder (1993) for more examples and details on these results.

In Figure 2.6 we show the implications of the above proposition for the binomial model when $T = 2$. We construct such a plot as follows. If we plot for each value of θ the vector for the binomial density function $\mathbf{f}_\theta = (\theta^2, 2\theta(1-\theta), (1-\theta)^2)'$, then we trace out a curve in $\mathbf{P}^2$ that we call the *unicomponent density curve* and denote by

$$\mathbf{C} = \{\mathbf{f}_\theta \colon \theta \in [0,1]\}.$$

[*Exercise*: Find the points in the figure corresponding to $\theta = 0$, 1 and 0.5.]

The curve $\mathbf{C}$ corresponds to all the possible density vectors obtainable under the unicomponent binomial model. As we have already learned, the set of all probability densities obtainable as mixtures of binomials is the convex hull of $\mathbf{C}$, denoted $\mathbf{M} = \text{conv}(\mathbf{C})$. If we consider any fixed set of three binomial densities, we can see pictorially that they cannot fall on a single line, so that the mixture set determined by any three distinct θ gives identifiable parameters π.

2.4.3. *Full identifiability of m components.* We now raise the level of difficulty by considering the more statistically important question: If we allow the location of the support points ξ of Q to be unknown, but restrict their number to be no more than m, when will both sets of parameters π and ξ be identifiable? This difficult question has a very simple and elegant solution in the case of the one parameter exponential family. To motivate the answer, we return to Figure 2.6. Geometrically, we can consider three distinct cases, depending on where the true mixture density vector $\mathbf{f}_Q$ is located.

- Suppose that $m = 1$. Notice that in this case the density vector $\mathbf{f}_Q$ must be an element of $\mathbf{C}$. This class of densities must be identifiable since they are just the (identifiable) binomial densities. Hence $m = 1$ implies identifiability.

- Suppose $\mathbf{f}_Q \in \text{int}(\mathbf{M})$, the *interior* of $\mathbf{M}$ (relative to the simplex). Note that one can draw infinitely many lines that pass through $\mathbf{f}_Q$ and connect two points on $\mathbf{C}$. The intersection points correspond to two parameter values, say ϕ_1 and ϕ_2, and we can correspondingly write

$$\mathbf{f}_Q = \pi \mathbf{f}_{\phi_1} + (1 - \pi)\mathbf{f}_{\phi_2},$$

  for some π for every such pair. We can draw two conclusions: first, that every interior density has a representation with $m = 2$, but second, that no such two-component mixture can be identifiable.
- If $m = 2$ and $\mathbf{f}_Q \in \text{bdry}(\mathbf{M})$, the boundary of $\mathbf{M}$, then $\mathbf{f}_Q$ must lie on the bottom line of the triangle, where it is clear that it can be represented by a convex combination of the component density vectors corresponding to the two extreme values of θ, namely, 0 and 1. Moreover, this is a unique representation with two points.

Thus we have a situation in which $m = 1$ implies identifiability, but only some two-component mixtures are identified. If we change our method of counting support points somewhat, however, we can report a very simple rule for identifiability of the latent distribution.

Thus we define index(Q) to be the number of support points in the latent distribution Q, with the *special rule* that the support points at the left and right extremes of the parameter space be counted as 1/2 a support point. Thus, for example, in the above binomial example, a mixture of $\text{Bin}(2, 0)$ and $\text{Bin}(2, 0.5)$ corresponds to a latent distribution with index 1.5, and the identifiable mixtures on the bottom edge of the triangle have index 1.

With this said, we can summarize the binomial example by saying that the distribution Q can be determined from $\mathbf{f}_Q$ if and only if index$(Q) \leq 1$. This result can be generalized, but we first develop some important tools in the following subsection.

2.4.4. *Hyperplanes and convex sets.* A useful tool for working with convex sets is the idea of the support hyperplane. For every vector of length 1, say $\mathbf{w}$, in $\mathbf{R}^d$, and every constant c, there exists a *hyperplane* $\mathbf{H} = \mathbf{H}(\mathbf{w}, c)$, defined to be the set

$$\mathbf{H} = \{\mathbf{v} \in \mathbf{R}^d\colon\ \mathbf{v}'\mathbf{w} = c\}.$$

It is a $(d - 1)$-dimensional linear surface. We can think of it as a translation of the linear subspace consisting of all vectors orthogonal to $\mathbf{w}$, namely,

$$\mathbf{H}(\mathbf{w}, c) = c\mathbf{w} + \langle \mathbf{w} \rangle^{\perp}.$$

[*Exercise.*]

In $\mathbf{R}^2$, a hyperplane is just a line in the plane. In $\mathbf{R}^3$ it is a two-dimensional planar surface. Each hyperplane can be associated with a *lower half space* $\{\mathbf{v} \in \mathbf{R}^d\colon\ \mathbf{v}'\mathbf{w} \leq c\}$ and an *upper half space* $\{\mathbf{v} \in \mathbf{R}^d\colon\ \mathbf{v}'\mathbf{w} \geq c\}$.

A *support hyperplane* to a convex set $\mathbf{B}$ is a hyperplane that bounds the set on some side. More formally, for each direction vector $\mathbf{w}$, let

$$c^*(\mathbf{w}) = \sup\{\mathbf{w}'\mathbf{b}:\ \mathbf{b} \in \mathbf{B}\}.$$

Verify that for every $c < c^*$ the convex set $\mathbf{B}$ must have a *nonempty* intersection with the upper half space of $\mathbf{H}(\mathbf{w}, c)$. For every $c > c^*$, the convex set $\mathbf{B}$ must have an *empty* intersection with the upper half space, and so it is contained in the lower half space. In this case, $\mathbf{H}(\mathbf{w}, c^*(\mathbf{w}))$ is a support hyperplane.

If the convex set $\mathbf{B}$ is closed, then it is clear that the support hyperplane $\mathbf{H}$ intersects $\mathbf{B}$ along a boundary and that the intersection consists of those points $\mathbf{b} \in \mathbf{B}$ satisfying $\mathbf{w}'\mathbf{b} = c^*$, whereas $\mathbf{w}'\mathbf{b} < c^*$ for all other $\mathbf{b} \in \mathbf{B}$. Thus the closed convex set lies completely in one of the half spaces generated by the hyperplane, with some of its boundary points in the hyperplane. (Indeed, a closed convex set can be represented as the intersection of the lower half spaces of its support hyperplanes.)

As an aside, we note that if $\mathbf{B}$ is a convex set, then $\mathbf{B}^* = \{\mathbf{v}:\ \mathbf{v} \cdot \mathbf{b} \le 1$ for all $\mathbf{b} \in \mathbf{B}\}$ is a *dual* convex set to $\mathbf{B}$. Note that if $\mathbf{b}^* \in \mathbf{B}^*$, then the set $\mathbf{B}$ is in the lower half space $\{\mathbf{u}:\ \mathbf{u} \cdot \mathbf{b}^* \le 1\}$. It follows that if $\mathbf{B}$ is closed, then $\mathbf{B}^*$ is closed and its boundary points correspond to the support hyperplanes of the set $\mathbf{B}$.

When we are working with a convex set $\mathbf{B}$ in the probability simplex, the entire set lies within the hyperplane $\mathbf{H}(\mathbf{1}, 1)$. Thus this hyperplane is a support hyperplane, but not a very interesting one as far as describing the set $\mathbf{B}$. If $\mathbf{w} \neq \mathbf{1}$, then a hyperplane $\mathbf{H}(\mathbf{w}, c)$ will intersect $\mathbf{H}(\mathbf{1}, 1)$ in a linear manifold of dimension $d-2$, and there will be multiple hyperplanes $\mathbf{H}(\mathbf{w}, c)$ that generate the same manifold within the simplex.

In particular, if $\mathbf{p}$ is in $\mathbf{P}^T$ and lies in the hyperplane $\sum w_j p_j = c$ determined by $\mathbf{H}(\mathbf{w}, c)$, then, since $\sum p_i = 1$, it also lies in the hyperplane $\sum (w_j - c) p_j = 0$, which is created by the hyperplane $\mathbf{H}(\mathbf{w} - c\mathbf{1}, 0)$.

The fact that in the probability simplex, and therefore in the mixture problem, one can reduce attention to hyperplanes with $c = 0$, and therefore containing the origin, turns out to be quite important in reducing the dimensionality of the mixture problem.

2.4.5. *Identifiability of weights and supports.* We now use the tools of support hyperplanes to turn questions about the structure of the boundary of a mixture set into questions about polynomials and so solve identifiability questions.

PROPOSITION 6. *If $f(x; \phi)$ is a discrete exponential family density (more generally a Chebyshev system density) with $T+1$ points of support, then the class of identifiable mixtures is those in the boundary of $\mathbf{M}$, which is exactly those elements satisfying*

$$\operatorname{index}(Q) \le T/2.$$

PROOF. For the proof here we simply use the binomial model, because the results can then be derived using well known results regarding polynomials. We start by showing that boundary points have the specified bound on their index. Suppose point $\mathbf{f}$ is in the boundary of the mixture density set $\mathbf{M}$ and so lies in a support hyperplane $\mathbf{H}$ defined by

$$\mathbf{w}'\mathbf{m} \leq 0 \quad \text{for all } \mathbf{m} \in \mathbf{M},$$

with $\mathbf{w}'\mathbf{f} = 0$. It follows that $\mathbf{f}$ can be represented as a mixture of the binomial vectors $\mathbf{f}_\theta$ that lie in that same hyperplane, hence satisfying $\mathbf{w}'\mathbf{f}_\theta = 0$; otherwise, one could use the mixture representation to show $\mathbf{w}'\mathbf{f} < 0$. However, the function

$$g(\theta) := \mathbf{w}'\mathbf{f}_\theta$$

is a polynomial in θ of degree T, so it has at most T roots. Moreover, since we are in a support hyperplane, with $\mathbf{w}'\mathbf{f}_\theta \leq 0$ for all θ, these roots must correspond to local maxima of g. Hence any root in the open interval $(0, 1)$ must be a root of even multiplicity to the polynomial. Since each root corresponds exactly to a potential support point, this means that if we count each support point in $(0, 1)$ with weight 2 and mass points at the extremal values of 0 and 1 with weight 1, then the total cannot exceed T. We can now apply the definition of index to argue that the maximal index of the latent distribution corresponding to $\mathbf{f}$ is $T/2$. Moreover, since the roots correspond to all of the possible support points, the preceding proposition shows that this latent distribution is unique and so is identifiable. To show that every mixture of index no more than $T/2$ is in the boundary, we can construct a polynomial with the roots corresponding to the support points and show that this polynomial implies the existence of a bounding hyperplane. On the other hand, if $\mathbf{f}$ is in the interior of $\mathbf{M}$, then we can again argue that we can draw many lines through it connecting two boundary points, each of which corresponds to a different mixture representation, and so nonidentifiability holds. □

As an exercise, the reader should consider how the proof might extend from the binomial to the case of the Chebyshev system.

More precise descriptions of elements of the interior of $\mathbf{M}$ are available, including the existence of exactly two representations of $\mathbf{f}$ in terms of mixtures of index $(T+1)/2$. If T is even, then there is one representation involving each of the two extreme parameter values [check this pictorially for $\text{Bin}(2, \theta)$]. If T is odd, then there is one representation in terms of $(T + 1)/2$ interior components and one involving $(T - 1)/2$ interior components and the two extreme components. From a statistical point of view, the latter representation has one more component than the former.

Fortunately, there is a simple rule of thumb that describes the identifiability of the parameters in a finite mixture model. If you specify a mixture model with m interior components and if the total number of free parameters $2m - 1$ is less than or equal to the number of free parameters in the multinomial, here T, then in fact the parameters are identifiable.

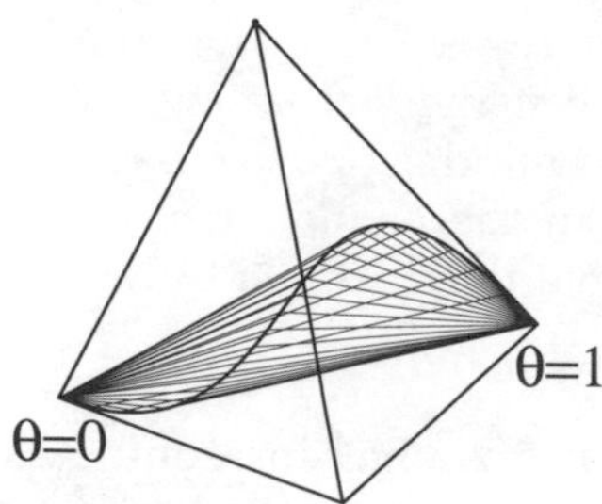

FIG. 2.7. *The unicomponent density curve for the* $\text{Bin}(3, \theta)$ *model.*

The following example illustrates the geometric structures corresponding to this index result. We consider the $\text{Bin}(3, \theta)$ model, with sample space $\{0, 1, 2, 3\}$. The probability simplex $\mathbf{P}^3$ can be represented as a tetrahedron, shown in Figure 2.7.

The curve $\mathbf{C} = \{\mathbf{f}_\theta\}$ starts at $(1, 0, 0, 0)$, where $\theta = 0$, and moves sinuously around to $(0, 0, 0, 1)$, where $\theta = 1$. The mixture set $\mathbf{M}$ has two smooth two-dimensional boundary surfaces, corresponding to the two different types of mixtures of index 1.5. One boundary corresponds to latent distributions with two support points, one of which is $\theta = 0$; the other is two point mixtures with one mass point at $\theta = 1$. The one component curve $\mathbf{C}$ corresponds to one seam along these two surfaces; the other seam is along one edge of the tetrahedron and is formed by mixtures of index 1 that are mixtures with support at $\theta = 0$ and 1. These boundary mixtures are identifiable, whereas no interior points have identifiable latent distributions.

In Figure 2.8 we show a cross section through the set $\mathbf{M}$, where we cut through the tetrahedron in the plane of all density vectors $\mathbf{p}$ on $\{0, 1, 2, 3\}$ with mean 1.5; that is, $\sum p(x) \cdot x = 1.5$. The two edges of the set $\mathbf{M}$ are the boundaries corresponding to the mixtures with index 1.5.

2.4.6. *Related problems.* When one leaves the i.i.d. case and considers other independent but not identically distributed structures, the analysis can be considerably more difficult and the results less simple. See, for example, Follman and Lambert (1991). Another complication arises in the i.i.d. case if one has nonlatent parameters in the model, so that one must address joint identifiability. Lindsay, Clogg and Grego (1991) managed to solve one such joint identifiability question in the Rasch model.

We conclude by noting that there are many interesting relationships between binomial mixture models and other natural distributions generated by sequences of binary variables. We point out two such cases.

Consider the distribution $\text{Bin}(1, p_1) * \text{Bin}(1, p_2)$, the convolution of two Bernoulli trials with success probabilities p_1 and p_2, respectively. This is a distribution on $\{0, 1, 2\}$ and so has density vector in $\mathbf{P}^2$. When $p_1 = p_2 := p$, the distribution is $\text{Bin}(2, p)$. As an *exercise*, show that the set consisting of all

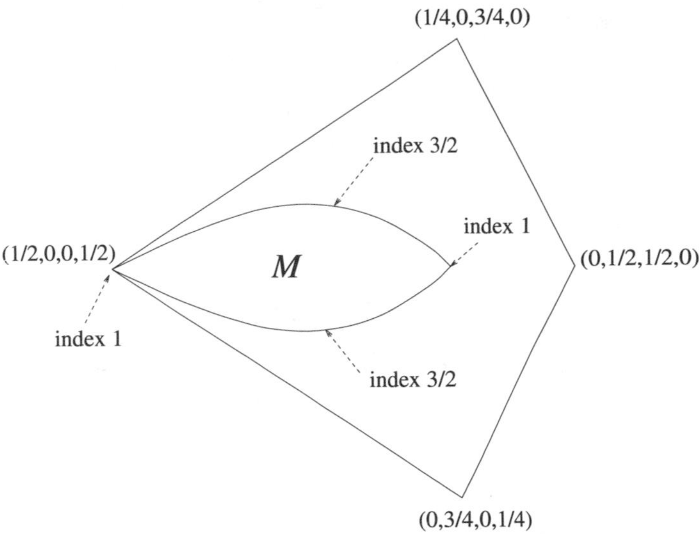

FIG. 2.8. *Cross section of the binomial mixture density set.*

Bernoulli convolutions with $p_1 \neq p_2$ is exactly the complement of the set of Bin(2, p) mixtures.

We can think of the variation of the value of p *between* the Bernoulli trials as being a distinct and nonoverlapping form of distributional assumption to the type of variation which occurs in the mixture model. There are a number of interesting papers on the structure of the space of Bernoulli convolutions, for example, Hoeffding (1956), but most pertinent for data analysis is that these models contrast with the mixtures in that the resulting convolution distributions are *lighter* tailed than the binomial distributions. For example, it can be shown that the binomial distribution is a dilation of any convolution distribution with the same mean.

Another interesting relationship arises from de Finetti's theorem. If we have an *infinite* sequence of exchangeable binary variates, say $X_1, X_2, \ldots$, then this theorem indicates that the sequence has a distribution that can be represented as a mixture, over latent variable p, of sequences of Bernoulli trials with success probability p. Diaconis (1977) investigated the implications of this result for *finite* sequences of exchangeable binary variates. These results indicate that if a finite sequence of binary variates is exchangeable, then the sum either has the distribution of a mixture of binomials or very nearly so, where Diaconis makes the "nearly so" statement precise.

2.5. Moment representations. We have now found out a great deal about the structure of the exponential family mixture, but a fundamental question remains: Given a density vector $\mathbf{p} \in \mathbf{P}^T$, is it possible to determine in a straightforward way whether $\mathbf{p}$ is in our mixture model? That is, whether or not $\mathbf{p} \in \mathbf{M}$. One simple test is already available to us from (2.4). That is, we can check to see if

$$\ln\left(\frac{p(t)}{f(t;\phi_0)}\right) \text{ is convex.}$$

If it is not, then **p** cannot be a set of mixture probabilities. Indeed, this kind of plot is very similar to ratio plots earlier in this chapter.

However, we can greatly sharpen this result in the special case when the density is exponential family on the lattice $\{0, 1, 2, \ldots, T\}$. Recall that in this circumstance, the ratios

$$R(t) = f(t; Q)/f(t; \phi_0) = \int \exp((\Phi - \phi_0)t)\, d\Gamma(\Phi)$$

are values of the moment generating function of some positive measure Γ. When the values of t are on the lattice, this implies that

$$R(0), R(1), \ldots, R(T)$$

are the moments of the nonnegative measure corresponding to the distribution of $\exp(\Phi - \phi_0)$. The argument can be used in reverse to show that **p** is a set of mixture probabilities if and only if

$$\frac{p(0)}{f(0;\phi_0)}, \ldots, \frac{p(T)}{f(T;\phi_0)}$$

are the moments of some measure on $(0, \infty)$. [For simplicity, we are here assuming the parameter space is $(-\infty, +\infty)$.]

Thus our question is equivalent to the following: When is a sequence of $T+1$ numbers, say $m_0, \ldots, m_T$, equal to the sequence of moments $\int x^k\, d\nu(x)$ for some positive measure ν with full mass on $(0, \infty)$?

The answer can be specified most easily through the use of moment matrices. We form a sequence of *moment matrices* M_p as follows. If p is even, say $p = 2k$, then

$$M_{2k} := \begin{bmatrix} m_0 & m_1 & m_2 & \cdots & m_k \\ m_1 & m_2 & m_3 & \cdots & m_{k+1} \\ m_2 & m_3 & m_4 & \cdots & \cdot \\ \vdots & \vdots & \vdots & \vdots & \vdots \\ m_k & m_{k+1} & \cdot & \cdot & m_{2k} \end{bmatrix}.$$

If p is odd, say $2k+1$, then

$$M_{2k+1} := \begin{bmatrix} m_1 & m_2 & m_3 & \cdots & m_{k+1} \\ m_2 & m_3 & m_4 & \cdots & \cdot \\ m_3 & m_4 & m_5 & \cdots & \cdot \\ \vdots & \vdots & \vdots & \vdots & \cdot \\ m_{k+1} & \cdot & \cdot & \cdot & m_{2k+1} \end{bmatrix}.$$

There are many results known about the nature of these matrices and their relationship to the number of mass points in a positive measure, with the special case of $m_0 = 1$ being that of a probability measure. For our purposes we note the following:

- If both M_{2k} and M_{2k-1} are positive definite, then $m_0, m_1, \ldots, m_{2k}$ are the moments for some nonnegative measure $d\gamma(\phi)$ on $[0, \infty)$. (In fact, there exist infinitely many measures with these moments.)
- If $m_0, m_1, \ldots, m_{2p}, \ldots$ are the moments of a measure with exactly K support points, then the sequence of matrices $M_0, M_1, M_2, \ldots, M_p, \ldots$ has the property that they are positive definite for $p \le 2K-2$ and are nonnegative definite with rank K, thereafter.

These results enable us to test if a set of points is a moment sequence. In particular, if we are testing if a sequence of ratios $R(t)$ is in the interior of the mixture space, then the fact that they lie in a region of nonidentifiable distributions means that the highest order moment matrices that can be constructed from the sequence must be strictly positive definite. To show that a sequence of ratios is in the boundary of the mixture space corresponding to some K point mixture, it suffices to show that the matrices are nonnegative definite, with rank K. (Note that the entire moment sequence is determined uniquely after maximal rank is reached, using the fact that the determinant of a rank deficient matrix is zero.)

This approach to testing for the presence of mixture structure was discussed in Lindsay, Clogg and Grego (1991).

2.6. Certain nested mixture models. Many times we will wish to apply the theory of mixtures to models that have auxiliary (nonlatent) parameters θ and so fall into the class of semiparametric mixture models. An important statistical question, about which we know relatively little in general, is the nature of the identifiability of the auxiliary parameters in the presence of the latent distribution. However, there are certain important cases, including the following normal example, in which it is indisputable that a fundamental loss of identifiability occurs and that we must be aware of the consequences of this loss. In particular, when it occurs, it may be impossible to sensibly estimate the parameters in the presence of the nonidentifiability.

As a first example, we consider the *normal mixture* problem. We assume that the component densities are $N(\mu, \sigma^2)$ and that there is an unknown latent distribution Q on the mean parameter μ, together with an unknown variance parameter σ^2 that is common to all the component densities. We write this model as $N(Q, \sigma^2)$.

We first prove that if μ has a latent distribution that is $N(\alpha, \tau^2)$, then the marginal distribution of X is $N(\alpha, \sigma^2 + \tau^2)$. Notice that in this convolution model, $X =_{\text{dist}} \Phi + Z$, where Φ and Z are independent normal variables with means α and 0 and variances τ^2 and σ^2, respectively, so this claim is just a standard result about the convolution of two normals, provable using moment generating functions:

$$\begin{aligned}E(\exp(tX)) &= E\left(\exp(t(\Phi + Z))\right) = E(\exp(t\Phi))E(\exp(tZ)) \\ &= \exp\left(\alpha t + \frac{\tau^2 t^2}{2} + \frac{\sigma^2 t^2}{2}\right).\end{aligned}$$

It follows that any mixture $N(Q, \sigma^2)$ can *also* be represented as a normal mixture by $N(Q^*, \sigma^2 - \delta)$, where Q^* is the convolution of Q and $N(0, \delta)$. Thus the class of mixture distributions, as σ varies, are *nested*,

$$\{N(Q, \sigma^2 + \tau^2)\colon\ Q \in \text{p.m.}\} \subset \{N(Q; \sigma^2)\colon\ Q \in \text{p.m.}\}$$

and the joint parameters (Q, σ) are *not* identifiable.

We do note that if we restrict attention to finite *discrete* latent distributions Q, then the pair (Q, σ) is identifiable. This can be shown by the fact that if Q has p points of support, then the pair can be recovered from the first $2p$ moments of X using the methods of Lindsay (1989b).

One might ask what happens to the nonparametric maximum likelihood approach to estimating (Q, σ). It is easily seen that there exists a nonparametric MLE for Q, say $\hat{Q}_\sigma$, for each fixed value of σ, because the likelihood for a sample is then bounded. However, the profile likelihood function $L(\hat{Q}_\sigma, \sigma)$ is clearly decreasing in σ by the above nesting property, since increasing the value of σ shrinks the class of eligible models that can be maximized over. In fact, as σ gets small, the solution $\hat{Q}_\sigma$ converges weakly to the empirical distribution for the data, and the profile likelihood becomes infinite. See, for example, Hathaway (1985).

Although this means that the likelihood method fails in this example, we do note that this is related to the fact that the class of normal mixtures is very flexible, providing smooth approximations to many other distributions. Roeder (1990) exploited this feature to generate a method of density estimation based on using estimators of the form $N(Q, \sigma^2)$, with parameter selection based on a goodness-of-fit criterion.

There are other two parameter exponential families that have a nested mixing structure. Jewell (1982) showed that the Weibull mixture families had such a structure, with the shape parameter playing the role of σ. As an illustration of his approach, we show how the technique is readily extended to the gamma family.

Consider the two parameter gamma density

$$f(x; \alpha, \lambda) = \frac{\lambda^\alpha}{\Gamma(\alpha)} x^{\alpha-1} e^{-\lambda x} \mathscr{I}[x > 0].$$

We ask if for any given α and λ_0, and any δ, with $\alpha - \delta > 0$, there exists a latent distribution Q such that

$$\int f(x; \alpha, \lambda)\, dQ(\lambda) = f(x; \alpha - \delta, \lambda_0). \tag{2.9}$$

We can write this equation symbolically as

$$\text{Gam}(\alpha, Q) = \text{Gam}(\alpha - \delta, \lambda_0).$$

This equation can be rearranged so that we seek a solution Q to

$$\int \lambda^\alpha \exp(-\lambda x)\, dQ(\lambda) = x^{-\delta} \exp(-\lambda_0 x) c,$$

where c is a finite constant.

First we ask if we should expect a solution. Recall that a necessary condition for a density $p(x)$ to be a mixture of a one parameter exponential family $\exp(\phi x - \kappa(\phi))$ is that $\ln(p(x)/f(x;\phi))$ is convex. Note that convexity holds in this example because $-\delta \ln(x)$ is convex in x. To go further, it is natural to exploit certain results on Laplace transforms, such as found in Feller [(1971, Vol. II, pages 439–441)]. A function $f(x)$ is said to be *completely monotone* on $[0,\infty)$ if its derivatives have the alternating sign property $(-1)^n f^{(n)}(x) \geq 0$. It is known that the function f is completely monotone if and only if it is the Laplace transform of a measure μ on $[0,\infty)$; that is, $f(x) = \int e^{-\lambda x}\, d\mu(\lambda)$.

It is easy to see that the function $x^{-\delta} e^{-x\lambda_0}$ is the product of two completely monotone functions and as such, it must be completely monotone. Let μ be its generating measure. We then have, if we set $dQ(\lambda) = c\lambda^{-\alpha}\, d\mu(\lambda)$, a formal solution to our problem. The one remaining point to check is that the measure so generated is finite, in the sense that $\int \lambda^{-\alpha}\, d\mu(\lambda) < \infty$. This can be checked as follows: Since

$$\int x^{\alpha-1} \exp(-\lambda x)\, d\mu(\lambda) = x^{\alpha-1-\delta} \exp(-\lambda_0 x),$$

we integrate both sides over x in $[0,\infty)$. The left-hand side is proportional to $\int \lambda^{-\alpha}\, d\mu(\lambda)$ and the right-hand side has a finite integral provided that $\alpha - \delta > 0$, which we have assumed.

Now that (2.9) is established, it follows that the gamma mixture models are also nested:

PROPOSITION 7. $\{\mathrm{Gam}(\alpha+\delta, Q)\} \subset_{\cdot} \{\mathrm{Gam}(\alpha, Q)\}$.

[*Exercise*: As a more direct proof, show that the latent distribution Q with density proportional to $(\lambda - \lambda_0)^{\delta}\, \lambda^{-\alpha}\, \mathscr{I}[\lambda > \lambda_0]\, d\lambda$ does the job in (2.9).]

2.7. Concluding remark. This chapter has given an introduction to some of the key ideas regarding the mathematical structures of mixture models. For much more on these properties, see the papers referenced in the text. For more general information on the identifiability question, in addition to the standard books on mixture models, there is the recent book by Prakasa Rao (1992), that has a chapter devoted to the identifiability of mixtures.

CHAPTER 3

Parametric Models

As an aid to understanding the role of nonparametric maximum likelihood, this chapter is a review of some of the basic features of the standard parametric models used in mixture modeling, together with basic features of maximum likelihood estimation in these models. We will consider the two different schools of modeling. In one, the latent distributions Q are assumed discrete, with a fixed number of components m. In the other, the latent distributions is assumed to come from some parametric family of continuous distributions.

3.1. Discrete versus continuous. There are certainly instances in which the latent distribution is logically and naturally modeled in either the discrete or continuous form due to the nature of the application. For example, it might be known that there are a finite and known number m of physical components in the population or, alternatively, that there is an inherent continuity expected in the latent variate Φ, such as when it represents a continuous covariate that was not measured.

More typically, however, we are on uncertain ground in specifying the number of components in a discrete latent distribution. Alternatively, in the case of a continuous variate Φ, we have no reason to believe it follows any particular parametric form. Although it is standard practice to assume normality for the latent variate, there is an element of arbitrariness in choosing the appropriate function of the latent variate to be normally distributed.

Moreover, as we have indicated already, we generally obtain very little goodness-of-fit information about the latent distribution from the data, and so there is little hope of having a large enough sample to investigate the true distributional form.

My own preference is for the use of discrete distributions in cases of doubt. The argument is that it makes maximum likelihood numerically simpler, and so is to be preferred if the form of the latent distribution specification makes little difference to the desired statistical inference. A number of investigations have found that misspecifying the latent distribution has very little effect on bias, and minimal effect on standard errors beyond the necessary correction for overdispersion when mixing is present [Butler and Louis (1992); Neuhaus, Hauk and Kalbfleisch (1992)].

One desirable feature of using a continuous family of latent distributions is that it leads to a smooth family of posterior densities $q(\phi|x)$ for empirical Bayes inference about the latent variates. However, this comes at the risk of allowing the user to imagine there is fairly precise information about the shape of the posterior, even though it is not possible to verify this shape empirically without gargantuan sample sizes.

On the behalf of using discrete mixtures, I would further argue that there is a much greater degree of flexibility—admittedly at the cost of additional parameters—without any real loss in fitting power. Moreover, as we will see, it is generally much simpler and more reliable numerically to calculate the estimates. In particular, a key limitation of the continuous approach is that, except for limited cases, the calculation of the mixture density $\int f(x;\phi)\cdot q(\phi)\,d\phi$ will require numerical integration, and unlike the discrete case, one usually cannot directly apply the reliable EM algorithm to perform the steps in the likelihood maximization.

3.1.1. *Continuous models: The conjugate family.* Although there are many instances where normality is assumed for the distribution of Φ, there is a more sophisticated approach to the construction of the latent distribution that exploits certain Bayesian structures, where the latent distribution plays the role of *prior*.

If a parametric family of prior densities $q(\phi;a,b)$ is such that the posterior densities $q(\phi|x;a,b)$ are from the same parametric family, then the family is said to be *conjugate* to the sampling density $f(x;\phi) = f(x|\phi)$. Such a conjugacy relationship holds, for example, if f and q are normal densities, because then the posterior densities are also normal.

Diaconis and Ylvisaker (1979) worked out a number of the properties for conjugate families in the case when f is from the exponential family. Suppose the density for a single observation is from the one parameter exponential family considered in the previous chapter, having density

$$f(x;\phi) = \exp(\phi x - \kappa(\phi))$$

with respect to some measure $dF_0(x)$. If we obtain a sample $X_1,\dots,X_n$, then the joint density has the form

$$f(\mathbf{x};\phi) = \exp(\phi(n\overline{x}) - n\kappa(\phi))$$

with respect to the product measure, where $\bar{x} = n^{-1}\sum x_i$.

Suppose that the latent parameter Φ has a continuous prior density from the two parameter conjugate family of models:

$$dQ(\phi) = q(\phi;\mu,\tilde{n})\,d\phi = \exp(\tilde{n}\mu\phi - \tilde{n}\kappa(\phi) - \psi(\mu,\tilde{n}))\,d\phi.$$

[As an *exercise*, show that the conjugate family for the binomial distribution is the Beta(α,β) distribution, with suitable selection of parameters.] It follows by examination of the joint density $f(x;\phi)\cdot q(\phi)$ that the posterior density $q(\phi|x)$ must be from the same family, with updated parameters

$$\mu^* = \alpha\mu + (1-\alpha)\overline{x},$$

with $\alpha := \tilde{n}/(n+\tilde{n})$, and

$$\tilde{n}^* = \tilde{n} + n.$$

[*Exercise.*] The form of the updating suggests a natural interpretation of the parameter $\tilde{n}$: it is the prior "sample size". Changing it by one unit has exactly the same effect on the posterior as does changing the sample size n by one unit. Additionally, the posterior parameter μ^* is very elegantly expressed as a weighted mean of the prior parameter μ and the sample mean $\overline{x}$, with weights proportional to the "prior sample size" $\tilde{n}$ and the sample size n.

Some useful characterizations of the moments of $\overline{X}$ can be obtained by using integration-by-parts techniques. If $h(\phi)$ is a differentiable function and $q(\phi) = q(\phi;\mu,\tilde{n})$, then we can easily show

$$[h(\phi)\cdot q(\phi)]' = h'(\phi)\cdot q(\phi) + h(\phi)\cdot[\tilde{n}\mu - \tilde{n}\kappa'(\phi)]\cdot q(\phi).$$

Suppose that $h(\phi)q(\phi)$ is zero at the left and right limits of the parameter space. Then integration of the last displayed equation gives us

$$0 = E[h'(\Phi)] + \tilde{n}\cdot E[h(\Phi)\cdot\{\mu - \kappa'(\Phi)\}]. \tag{3.1}$$

Replacing $h(\phi)$ in (3.1) with various functions of interest now gives a number of useful identities regarding the marginal distribution of the observable variable X. If we let $h(\phi) = 1$, then it proves that

$$\mu = E[\kappa'(\Phi)].$$

That is, in terms of the prior, or latent, distribution, μ has a natural interpretation. This can be turned into a property for the mean of X by recalling that in a one parameter exponential family,

$$\kappa'(\phi) = E[X;\phi].$$

Hence in terms of the sampling distribution for X, the parameter μ equals $E[\ E[X|\Phi]\ ] = E[X]$, the marginal mean of X.

Applying the integration-by-parts trick to $h(\phi) = (\mu - \kappa'(\phi))$ yields a second identity

$$E[\kappa''(\Phi)] = \tilde{n}E\{\mu - \kappa'(\Phi)\}^2 = \tilde{n}\,\mathrm{Var}(\kappa'(\Phi)).$$

Since the left-hand side of the last equation is also $nE[\mathrm{Var}(\overline{X}|\Phi)]$, we have that

$$\mathrm{Var}(\overline{X}) = E[\mathrm{Var}(\overline{X}|\Phi)] + \mathrm{Var}[E(\overline{X}|\Phi)] = \mathrm{Var}(\kappa'(\Phi))\cdot\left[\frac{\tilde{n}}{n}+1\right].$$

Thus the sampling variance of the sample mean is an inflated version of the variability of the latent mean value parameter. [As an *exercise* in understanding, check that these formulas apply in the beta binomial example.]

The simple prior-to-posterior relationship makes the conjugate families very attractive in the Bayesian framework. They would also seem to have nice mathematical structures for mixture model inference. However, there are two additional points to consider. One is that the families have been constructed purely based on mathematical convenience, not modeling considerations, and so might leave the user with a goodness-of-fit problem. One solution to this is to enlarge the latent distribution model to allow finite mixtures of conjugate distributions, say mixed over the mean parameter μ, with the dispersion parameter $\tilde{n}$ fixed. Dalal and Hall (1983) show that this class is extremely rich, in that one can approximate arbitrarily closely any prior Q.

Another point to consider is that the tidy prior–posterior relationship provides no guarantees that the marginal mixture density has any nice features. Typically the resulting marginal distribution for X is no longer in the exponential family and so lacks the nice features thereof, such as reduction to sufficient statistics, guaranteed-to-be-unimodal likelihoods and uniformly most powerful test procedures. It is easily checked that the likelihood equations have the form

$$n^{-1}\sum E[\Phi|X = x_i] = E[\Phi],$$
$$n^{-1}\sum E[\kappa(\Phi)|X = x_i] = E[\kappa(\Phi)].$$

Thus in the conjugate family model, $\hat{\mu} = \overline{x}$ does not seem to be implied by the likelihood equations. This occurs in contrast to the finite mixture model, where we will show that the likelihood equations imply that the sample mean is equated to the theoretical mean.

3.2. Discrete latent distribution. We now turn to the alternative form of parametric modeling, in which one assumes that the latent distribution is discrete, with a known number of components m. We will denote a latent distribution of this type as Q_m.

3.2.1. *Known component distributions.* We return to the models of Section 1.3.1, in which the mixture density can be written as $f(x; Q_m) = \sum f_j(x)\pi_j$, with known component densities f_j, $j = 1, \dots, m$. We now wish to set up the likelihood equations for the weight parameters. One obstacle is the need to satisfy the constraint $\sum \pi_j = 1$. One solution is to use a Lagrange multiplier technique. Another one is to eliminate one of the weights, say by setting $\pi_m = 1 - \pi_1 - \cdots - \pi_{m-1}$. If this is done, there are then $m - 1$ score functions

$$S_j(\boldsymbol{\pi}) = \frac{\partial \ln(L(Q_m))}{\partial \pi_j} = \sum \frac{f_j(x_i) - f_m(x_i)}{f(x_i; Q_m)}. \tag{3.2}$$

Unfortunately, the inequality constraints $\pi_j \geq 0$ mean that unless the maximum likelihood estimator $\hat{\pi}_j$ is in the interior of the parameter space, with all $\hat{\pi}_j$ strictly positive, the estimate does not solve the usual likelihood equations $S_j(\boldsymbol{\pi}) = 0$. Although we ignore this difficulty in this chapter, it should be pointed out that the nonparametric mixture maximum likelihood theorem of Chapter 1 does apply, and so the solution is completely described by the gradient inequality.

However, there are two pieces of good news. First, the log likelihood is *strictly concave* in the parameters π. [*Exercise*: Show that the matrix of second derivatives is negative definite.] This means that we have unique solutions for the weights, provided that they are identifiable. The second piece of good news is that the EM algorithm is easy to construct and implement.

Here is one derivation of the EM algorithm for this problem. By rearranging the likelihood equations $S_j(\boldsymbol{\pi}) = 0$, one obtains the equivalent equations

$$\pi_j = \pi_j \cdot n^{-1} \sum_{i=1}^{n} \frac{f_j(x_i)}{f(x_i; Q_m)}. \tag{3.3}$$

[*Exercise*.] This is a vector equation of the "fixed point" type, namely, $\boldsymbol{\pi} = \mathbf{F}(\boldsymbol{\pi})$, and the corresponding fixed point algorithm is simple: Given the current value $\boldsymbol{\pi}_c$, the updated value is $\boldsymbol{\pi}_u = \mathbf{F}(\boldsymbol{\pi}_c)$.

We will return to this algorithm later in order to give its missing data interpretation and some further background.

3.2.2. *Unknown component parameters.* If the m-component densities have unknown latent parameters $\xi_1, \ldots, \xi_m$, then we must also maximize the likelihood over them. Thus, in addition to the previous score functions for the weights, we will need the ξ scores:

$$U_j(\boldsymbol{\xi}, \boldsymbol{\pi}) = \frac{\partial}{\partial \xi_j} \ln(L(Q_m)) = \sum_i \left(\frac{f(x_i; \xi_j)}{f(x_i; Q_m)} \right) \left(\frac{f'(x_i; \xi_j)}{f(x_i; \xi_j)} \right). \tag{3.4}$$

We have written the scores in this fashion to emphasize that they are weighted sums of the *score functions*

$$v(x_i; \xi) := \frac{f'(x_i; \xi)}{f(x_i; \xi)}$$

from the unicomponent model.

The full m-component likelihood equations then require solving for the full set of scores (3.2) and (3.4) to equal zero. Unfortunately, the likelihood equations no longer need have unique solutions. Indeed, experience suggests that they frequently have multiple solutions. We will discuss this point further when discussing the issue of initial values for algorithms. On the other hand, despite their apparent complexity, there is the easy-to-program and reliable EM algorithm to implement for their solution.

3.3. Properties of the m-component MLE. We suppose that we have an m-component mixture estimator $\hat{Q}_m$ that satisfies the likelihood equations satisfied above. There are useful ways to paraphrase these equations that give some further insights. The following results are from Lindsay (1981).

PROPOSITION 8. *For any function* $g(\phi)$ *and* $\hat{Q}_m$ *an m-component solution determined by setting the* π *scores in* (3.2) *equal to zero, the following self-consistency equations hold*:

$$n^{-1}\sum E[g(\Phi)|X = x_i; \hat{Q}_m] = E[g(\Phi); \hat{Q}_m].$$

The proof is left as an *exercise*. The preceding result applies to both the known and unknown component cases. The score equations from the unknown components can also be given a prior–posterior expression as follows:

PROPOSITION 9. *For any function* $h(\phi)$ *and any m-component solution found by setting the* ξ*-score functions in* (3.4) *to zero, we have*

$$n^{-1}\sum E[\, h(\Phi)v(\Phi; x_i)\, |X = x_i; \hat{Q}_m] = 0.$$

We may put these two sets of equations together in the case of the exponential family mixture to obtain the following first moment property.

PROPOSITION 10. *If the component density* $f(x;\phi)$ *is a one parameter exponential family, and* $\hat{Q}_m$ *satisfies the m-component likelihood equations, then* $E[X; \hat{Q}_m] = \bar{x}$.

This may be proved by using $h(\phi) = 1$ and $g(\phi) = \kappa'(\phi)$ in the two preceding propositions.

3.4. EM algorithm. The next task is to give a derivation of the EM algorithm for the m-component discrete mixture model.

3.4.1. *A description of the EM.* The EM algorithm requires a particular model structure. Suppose that we have a model with parameters η in which there is both observed data X and missing data J. We need to maximize the likelihood of the observed data X, call it $L_X(\eta)$, but the likelihood is difficult to maximize. However, we assume that if we knew the unobserved data J, then the maximization of the likelihood $L_{(X,J)}(\eta)$ of the pair (X, J) would be easy, ideally having explicit solutions. We note that the "missing data" J could be completely imaginary; the important thing is that the distribution of the variable X that is observed is exactly the same as the marginal distribution of X in some hypothetical pair (X, J) which has an "easier" likelihood.

The *E step* in the EM algorithm in this situation involves taking a current value η_c and finding the *EM log likelihood* l_{em}, which is the conditional expectation of the full data log likelihood $\ln(L_{(X,J)})$ given the observed data:

$$l_{\text{em}}(\eta; \eta_c) := E[\,\ln L_{(X,J)}(\eta) \mid X; \eta_c].$$

The *M step* in a cycle is to let the EM solution η_{c+1} be that value of η, that maximizes the EM likelihood $l_{\text{em}}(\eta; \eta_c)$. Ideally, both steps have explicit solutions.

It is a simple *exercise* in the use of Jensen's inequality to show that such an iterative sequence increases the true likelihood L_X at each cycle [Dempster, Laird and Rubin (1977)].

3.4.2. *The EM for finite mixtures.* We will set these equations up in the framework of the discrete component density. The advantage to this approach is that we can give a simple interpretation of the action of the algorithm in terms of the filling-in-by-expectation of the unobserved cells of a contingency table. The formulas are exactly the same in the continuous case, but this author finds the contingency table approach more insightful.

In the multinomial case we can reduce the observed data by sufficiency to the counts $n(t) = \#\{X_i = t\}$. The missing data will be the variables J_i that identify the component labels (introduced in Chapter 1, Section 1.1). Thus in a discrete setting, we can reduce the complete data by sufficiency to the counts $n_j(t)$ of the number of times the pair (t, j) appeared in the complete sample. The observed data are then the column marginal totals of an unobserved table of counts, where the *rows* correspond to the different components and the *columns* correspond to the possible values of the variable X: see Table 3.1. The corresponding table of multinomial cell probabilities is shown in Table 3.2.

The complete data log likelihood for such a multinomial model is simply

$$\sum_j \sum_t n_j(t) \ln[\pi_j f(t; \xi_j)] = \sum_j \ln(\pi_j)\Big(\sum_t n_j(t)\Big) + \sum_j \sum_t n_j(t) \ln(f(t; \xi_j)). \tag{3.5}$$

Since the entries $n_j(t)$ in the full table are unobserved, to calculate l_{em} we take the complete data log likelihood (3.5) and replace the $n_j(t)$ with their expectations, say $\tilde{n}_j(t)$, conditioned upon the observed data and the current parameter estimates. This is easily done by allocating the total observed counts in a column $n(t)$ to each row in that column proportionally to its current estimated cell probability, so that we have

$$\tilde{n}_j(t) = n(t) \cdot \frac{\pi_j f(t; \xi_j)}{f(t; Q_m)}.$$

Since the parameters π and ξ separate in the complete data likelihood (3.5), the EM algorithm updates the current estimates of the weights π by

TABLE 3.1

	$t = 0$	$t = 1$	$\cdots$	$t = T$
$J = 1$	$n_1(0)$	$n_1(1)$	$\cdots$	$n_1(T)$
$J = 2$	$n_2(0)$	$n_2(1)$	$\cdots$	$n_2(T)$
$\cdots$	$\cdots$	$\cdots$	$\cdots$	$\cdots$
$J = m$	$n_m(0)$	$n_m(1)$	$\cdots$	$n_m(T)$
Totals	$n(0)$	$n(1)$	$\cdots$	$n(T)$

TABLE 3.2

	$t=0$	$t=1$	...	$t=T$
$J=1$	$\pi_1 f(0;\xi_1)$	$\pi_1 f(1;\xi_1)$	$\cdots$	$\pi_1 f(T;\xi_1)$
$J=2$	$\pi_2 f(0;\xi_2)$	$\pi_2 f(1;\xi_2)$	$\cdots$	$\pi_2 f(T;\xi_2)$
...	...	...	...	...
$J=m$	$\pi_m f(0;\xi_m)$	$\pi_m f(1;\xi_m)$	$\cdots$	$\pi_m f(T;\xi_m)$
Totals	$f(0;Q)$	$f(1;Q)$	$\cdots$	$f(T;Q)$

maximizing the first summand of the likelihood above, yielding

$$\pi_{j,\text{em}} = \pi_j n^{-1} \sum_t \tilde{n}_j(t),$$

the same formula derived earlier as a fixed point algorithm.

Notice also the algorithm has a natural interpretation in terms of the posterior probabilities. The term

$$\pi_{j|i} := \Pr(J_i = j | X = x_i; Q_m)$$

represents the posterior probability that the ith observation was from component j. The updated estimate of the weight parameter equals the average of the current estimated posterior probabilities:

$$\pi_{j,\text{em}} = \Pr(J = j; \hat{Q}_{\text{em}}) = n^{-1} \sum_i \pi_{j|i}.$$

If the component parameters ξ_j are unknown, then we must add to the previous equations a set of equations for them. In this case, we must solve $\sum_t \tilde{n}_j(t) v(\xi_{j,\text{em}}; t) = 0$. This equation is related to the second proposition of the preceding section in that it specifies

$$\sum_i \pi_{j|i} v(\xi_{j,\text{em}}; x_i) = 0,$$

a weighted set of score equations. If the components have an exponential family density, then these equations specify that the EM iterations in the mean value parameters $\mu(\xi)$ are

$$\mu_{j,\text{em}} = \frac{\sum_i \pi_{j|i} \cdot x_i}{\sum_i \pi_{j|i}};$$

that is, they are simply reweighted means, where the weights are proportional to the posterior probability of having come from group j.

3.4.3. *Algorithmic theory.* A basic overview and some algorithmic theory for the EM algorithm in the mixture model can be found in Redner and Walker (1984). In addition to the simplicity of construction in the mixture model, which is very easy to program, it must be stated that it increases the likelihood at every step and so is quite reliable at finding local maxima of the likelihood.

However, one key feature of the algorithm is that it commonly displays a *very slow* linear rate of convergence, where the rate constant is related to the amount of information in the missing portion of the data. If the components are similar in their densities, then the convergence is extremely slow. The convergence will also be slow when the maximum likelihood solution requires some of the weight parameters to be zero, because the algorithm can never reach such a boundary point. Although this is not so severe a problem for computing point estimates with modern computers, it can make simulation studies quite tedious.

An additional and related problem is that of deciding when to stop the algorithm. One risk to the naive user is the natural tendency to use a *stopping rule* for the algorithm based on the changes in the parameters or the likelihood being sufficiently small. Unfortunately, taking small steps does not mean we are close to the solution. If we were to continue, we might end up taking *many* more steps of nearly the same size and arriving at the solution a long distance away.

To combat this problem, Böhning, Dietz, Schaub, Schlattman, and Lindsay (1994) exploit the regularity of the EM algorithmic process to predict, via the device known as *Aitken acceleration*, the value of the log likelihood at the maximum likelihood solution. This method is suitable whenever one is using a linearly convergent algorithm with a slow rate of convergence. If l_{i-2}, l_{i-1} and l_i are the log likelihood values for three consecutive steps of the algorithm, then the predicted final value is

$$l_i^\infty = l_{i-2} + \frac{1}{1-c_i}(l_{i-1} - l_{i-2}) \quad \text{where } c_i = \frac{l_i - l_{i-1}}{l_{i-1} - l_{i-2}}.$$

If the algorithm is moving slowly, then c_i, an estimate of the rate, will be close to 1, and l_i^∞ will be substantially larger than l_i.

Böhning, Dietz, Schaub, Schlattman and Lindsay (1994) used this device to carry out a simulation study of the likelihood ratio test for one component versus two components. It enabled them to predict the final maximized likelihood with many fewer iterations. The Aitken accelerated value can also be used to construct a stopping rule that more adequately captures the desired numerical accuracy than the usual "lack-of-progress" stopping rule criterion

$$\text{stop if } l_i - l_{i-1} < \text{tol},$$

where tol is a prespecified tolerance level. Provided that l_i^∞ is a good estimator of the final likelihood, the rule

$$\text{stop if } l_i^\infty - l_i < \text{tol}$$

will cause the algorithm to stop only when the solution is near, and tol will more meaningfully represent the actual accuracy attained.

The idea of Aitken acceleration can be applied to the entire vector of parameters to speed up the algorithm itself [Louis (1982)]. However, as the number of parameters grows, this becomes more difficult to implement and less reliable. Other devices for speeding up the algorithm can be found in a variety of papers in the literature, proving that this is considered a serious problem and that no solution yet is completely satisfactory. Another point regarding the EM algorithm for mixtures is that, in fact, there are sometimes other ways to specify the "missing data," so there is not a unique EM algorithm. There are situations, such as the known component model, where there exist EM algorithms that are strictly superior to the one presented here, although the one here is the simplest.

As a final point, we relate the EM algorithm to the properties of the gradient function. First, we can express the EM steps for the weights as a form of gradient projection:

$$\pi_{j,\mathrm{em}} = \pi_j[1 + n^{-1} D_{Q_m}(\xi_j)].$$

The likelihood equations for the support point ξ_j can be written as

$$D'_{Q_m}(\xi_j) = 0.$$

The EM algorithm leaves the support point fixed if this equation is satisfied and will move it in the direction of increasing gradient value when it is violated. [*Exercise.*]

3.5. Multimodality and starting values. The presence of significant multimodality in the finite component likelihood has a number of important consequences.

For one thing, the solution of the likelihood equations can depend greatly on the initial values for the algorithmic method chosen. In an example given in Böhning, Schlattman and Lindsay (1992), it was found that in a particular problem in which the global maximum likelihood estimator for the two-component model had support points -1.6 and -6, starting the iterations of the EM algorithm with mass 0.5 at the points 0 and one of $-1, -2, -3, -4$ or -5 led to a second local maximum that had support points 0 and -1.8.

A second consequence of multimodality is that Newtonian-type algorithmic methods for obtaining solutions can be very unstable [Finch, Mendell and Thode (1989)].

Yet another important implication, one that will arise later in the consideration of the likelihood ratio test, is that the results obtained from a simulation study can be highly dependent on the stopping rules and search strategies employed. Indeed, this can make it quite difficult to compare the results of simulation studies or assess their reliability.

It is widely considered desirable to find the global maximum to the likelihood. If this is the goal, then one can adopt a number of strategies. One approach, suggested in Böhning, Schlattman and Lindsay (1992), is to calculate the nonparametric maximum likelihood estimator of Q, which can be done unambiguously. If it has more components than desired, one can often reasonably choose a way to consolidate nearby support points in a way that the likelihood stays near its maximum value. The use of this method was illustrated in the aforementioned example, and it was found that the four point NPMLE gave a good prediction of the best two-component MLE.

On the other hand, if the NPMLE has fewer components than desired, then this indicates that it will be *impossible* to find a maximum likelihood estimator with the desired number of components. (In such a situation, the EM algorithm iterations will either slowly merge the support points ξ together or force some values of π toward zero, because estimators with fewer support points have higher likelihoods than those with more.)

We should also note that this strategy for constructing initial values is not possible in some mixture models, such as the normal with unknown variance, in which a NPMLE does not exist.

Finch, Mendell and Thode (1989) suggest a strategy of multiple random starts that enables one to make a probabilistic estimate of the number of unseen modes to the likelihood. Another strategy that is possible in the two-component univariate normal is to use a normal probability plot to estimate the means and variance (or variances) from the slope and intercepts in the two tails. [See, e.g., Titterington, Smith and Makov (1985), pages 58–60.] This last approach would not be effective in a simulation study, nor would the NPMLE approach, unless they were made more systematic.

A point that is seldom raised is that it is not clear that one should insist on finding the global maximum to the likelihood. For example, if one is estimating the parameters in the normal mixture

$$\pi n(x; \mu_1, \sigma_1^2) + \overline{\pi} n(x; \mu_2, \sigma_2^2),$$

then the global maximum to the likelihood is ∞. This can be established by letting $\mu_2 = x_1$, or any other observation, and letting σ_2^2 go to zero. Then the term in the likelihood corresponding to x_1 becomes infinite while the other terms stay bounded below [Kiefer and Wolfowitz (1956)]. However, it is known that there is a consistent sequence of roots to the likelihood equation [Kiefer (1978)].

For these reasons, this author has been interested in constructing reliable and consistent estimators of the mixture parameters by *moment methods* and using these as initial values in likelihood algorithms. It is quite clear that if there is a consistent sequence of roots to the likelihood equations and we start our algorithm at the moment estimators, which will be $\sqrt{n}$ consistent, then, at least asymptotically, we should be finding a root close to the true values of the parameters. Lindsay (1989a) has an extensive development of the moment estimators, showing how they can be constructed in many important exponential family models. Among their important features is that they are unique, so that one need not hunt for the best solution, and that they are often numerically simple to compute compared to the maximum likelihood estimators.

The use of moment estimators was investigated extensively by simulation in the normal mixture model by Furman and Lindsay (1994a, b). They found that the moment estimators had high initial likelihoods, generally higher than using the true parameter values of the simulation, and that the EM iterations that were started at the true values, thereby leading to the solution of the likelihood equations closest to the truth, almost always picked the same solution as the EM iterations started at the moment estimators.

Following this successful experiment, Lindsay and Basak (1993) developed moment methods for the multivariate normal problem, which was considerably more challenging, but still resulted in a successful and numerically fast way to construct initial values.

CHAPTER 4

Testing for Latent Structure

We consider two closely related hypothesis testing problems. In the test (*homogeneity* (against heterogeneity), we test

$$H\text{: one component} \quad \text{versus} \quad A_1\text{: any mixture.}$$

In testing for the *number of components*, the usual test compares

$$H\text{: one component} \quad \text{versus} \quad A_2\text{: two components.}$$

These two problems are clearly closely related, and it can be anticipated tha tests valid for one might also be applied to the other. As such, we consider th two problems together.

In a survey of the tests of homogeneity, DerSimonian (1989) found that mos of the available procedures were one of two types. The first, the $C(\alpha)$ test fo homogeneity, is simple and tractable, with a known limiting distribution an a local optimality property. Strictly speaking, it is designed for the alterna tive A_1.

The second, the likelihood ratio test for H versus A_2, has long been a enigma. In addition to the aforementioned problems with multiple likelihoo roots when the number of components is fixed, we have the problem that th standard likelihood ratio statistic has an unknown limiting distribution. Ir terest has persisted in this method, no doubt partially due to the mystery, bu also related to the phenomenal record of reliability for this testing procedur when used in other standard problems.

The goals of this chapter are to provide essential background on these prc cedures, plus provide some fundamentally new insights to both. We will star this chapter with the $C(\alpha)$ testing procedure. It is a simple and highly e fective method of testing for overdispersion, and is the method the autho would recommend for most practical situations. A new result is given for it optimality, which had been clouded by work of Moran (1973).

The second part of this chapter is devoted to an extensive treatment of th asymptotic distribution of the likelihood ratio test for one versus two com ponents. A complete description is given for a class of multinomial mixtur models, together with a simple approximation formula. As by-products of this analysis, we derive the appropriate score test for the two-component problem and give a new result for the one component versus nonparametric mixture model as well. These results require a careful, and somewhat difficult, geometric analysis related to the geometry given in Chapter 2.

4.1. Dispersion score tests. The most popular method for testing for overdispersion is Neyman's $C(\alpha)$ test for homogeneity, although it is not always identified by this name because the same test can be derived in a number of different ways. We will start by motivating the test statistic from the point of view of the gradient function used in nonparametric maximum likelihood.

4.1.1. *The dispersion score.* Recall from the mixture NPMLE theorem of Chapter 1 that the degenerate distribution $\Delta_{\hat{\phi}}$ is the NPMLE for Q if and only if

$$D_1(\phi) \leq 0 \quad \text{for all } \phi. \tag{4.1}$$

Here $D_1(\phi)$ is unicomponent gradient function $D_{\Delta_{\hat{\phi}}}(\phi)$. We note also that we have

$$D_1(\hat{\phi}) = 0,$$

$$D_1'(\hat{\phi}) = 0.$$

The first equation is by direct calculation and the second, where the prime is the derivative with respect to ϕ, holds because of its equivalence to the likelihood equation which defines $\hat{\phi}$. [*Exercise.*] Thus by Taylor expansion about $\hat{\phi}$ we have, locally,

$$D_1(\phi) \approx \tfrac{1}{2}(\phi - \hat{\phi})^2 D_1''(\hat{\phi}). \tag{4.2}$$

It follows that the unicomponent gradient inequality (4.1) is violated in a neighborhood of $\hat{\phi}$ if $D_1''(\hat{\phi}) > 0$, but if < 0, it must hold locally. That is,

$$D_1''(\hat{\phi}) > 0 \implies \Delta_{\hat{\phi}} \text{ is not NPMLE.} \tag{4.3}$$

Thus $D_1''(\hat{\phi})$ seems to be an important summary statistic for checking a unicomponent hypothesis against local violations of that hypothesis. We therefore investigate its form. We define the (*Neyman*) *dispersion score* function to be

$$v_2(\phi, x_i) := \frac{f''(x_i;\phi)}{f(x_i;\phi)} = v(\phi; x_i)^2 + v'(\phi; x_i).$$

By direct calculation,

$$D_1''(\hat{\phi}) = \sum_i v_2(\hat{\phi}; x_i).$$

If we further examine the form of v_2 in the exponential family, using the natural parameterization, we find that

$$v_2(\phi, x) = (x - \mu)^2 - \sigma^2,$$

where μ and σ^2 are the mean and variance of the statistic X. Since under the null hypothesis, $\hat{\mu}_{\text{mle}} = \bar{x}$, we can conclude that there are no local gradient violations of the unicomponent model if and only if the sample variance is smaller than the variance predicted under the one-component model:

$$D_1''(\hat{\phi}) = \sum_i (x_i - \bar{x})^2 - n\sigma^2(\hat{\phi}).$$

Moreover, $D_1''(\hat{\phi})/n$ can be seen to be a consistent estimator of the variance of the mean value parameter μ under the latent distribution Q. [*Exercise.*]

However, the relationship of this statistic to the gradient function also indicates it may have low power to detect the need for a second component ϕ that is *not near* to the unicomponent estimate $\hat{\phi}$, because the gradient at a distant ϕ is likely to be less well predicted by the second order Taylor series approximation (4.2). Indeed, as we will see, the likelihood ratio test does use the gradient information for ϕ away from $\hat{\phi}$.

4.1.2. *Neyman and Scott's $C(\alpha)$ test.* For a second development of this statistic, we turn to Neyman and Scott's original derivation (1966). Let G be a distribution for Θ that has mean 0 and variance 1 (we assume the parameterization has been chosen so as to make this possible). We construct a location-scale family of distributions for Φ, with parameters (a, b) by setting

$$\Phi =_{\text{dist}} a + b\Theta,$$

where Θ has distribution G. [The parameters (a, b) may need to be restricted to ensure that the distribution has all its mass in the parameter space for ϕ.] We note that as the parameter b goes to zero, the distribution converges to a degenerate distribution at parameter a. Thus testing

$$H\colon\ b = 0 \quad \text{versus} \quad A\colon\ b > 0$$

is a test of overdispersion.

We consider the construction of the locally most powerful score test for this problem, so as to obtain maximum local power. We treat a as fixed for the moment. We calculate the first derivative of the log likelihood with respect to b:

$$\frac{\partial}{\partial b} \ln \int f(x; a + bs)\, dG(s) := S_1(a, b) = \frac{\int s f'(x; a + bs)\, dG(s)}{\int f(x; a + bs)\, dG(s)}.$$

Since G has mean zero, as $b \to 0$ this score converges to

$$\left[\int s\, dG(s)\right] \cdot \frac{f'(x; a)}{f(x; a)} = 0$$

under appropriate regularity conditions.

Since the first derivative of the log likelihood becomes degenerate as we approach the null, the approximate form of the *Neyman–Pearson tests* in a neighborhood of $b = 0$ must be determined by the second derivative of the log likelihood. Differentiating once more, we obtain

$$\frac{\partial^2}{\partial b^2} \ln \int f(x; a + bs)\, dG(s) = \frac{\int s^2 f''(x; a + bs)\, dG(s)}{\int f(x; a + bs)\, dG(s)} - S_1^2(a, b).$$

Since $\int s^2 dG(s) = 1$ by assumption, as $b \to 0$ this score converges to the dispersion score

$$v_2(a; x) = \frac{f''(x; a)}{f(x; a)}.$$

For a *fixed* value of a, the locally most powerful test of our hypotheses is therefore to reject for large positive values of $\sum v_2(a; x_i)$.

[*Note:* Earlier workers parameterized the family of latent distributions by (a, σ^2), with $\sigma^2 = b^2$. If one does this, one finds that the limit of the *first* derivative with respect to σ^2 is equal to (through the use of l'Hôpital's rule) the above limiting *second* derivative with respect to b. We have taken the scale parameter approach here because of an important point to be touched upon later.]

To complete construction of the $C(\alpha)$ procedure, as described by Neyman (1959), we must compute a $\sqrt{n}$ consistent estimator of a under the null hypothesis. Since the null hypothesis is the unicomponent model, we may use the one-component MLE $\hat{\phi}$. In this case, the $C(\alpha)$ test statistic will be the statistic we earlier derived from the gradient,

$$V_2 := \sum v_2(\hat{\phi}; x_i) = D_1''(\hat{\phi}),$$

normalized so as to have asymptotic variance 1. Note that except for the estimated parameter, V_2 is an i.i.d. sum. The asymptotic variance of V_2 is $n \cdot \tau(\phi)$, where

$$\tau(\phi) = E[v_2(X)]^2 - \frac{E^2[v_2(X)v_1(X)]}{E[v_1(X)]^2}.$$

Note that the second term in this variance formula is necessary to allow for the estimation of ϕ in V_2. Thus $V_2/\sqrt{n\tau(\hat{\phi})}$ is an asymptotically standard normal test statistic and has certain local optimality properties described by Neyman.

We note that the $C(\alpha)$ test can be constructed using a different point estimator than the MLE $\hat{\phi}$, but we must replace v_2 with a *corrected* dispersion score $\tilde{v}_2 := v_2 - \rho v_1$, where

$$\rho = \frac{E[v_2(X)v_1(X)]}{E[v_1(X)]^2}.$$

The adjustment is necessary to correct for the estimation of the nuisance parameter under the null hypothesis, but if one uses the MLE of the nuisance parameter, the correction term is zero. The variance term $\tau(\phi)$ is in fact equal to $\mathrm{Var}(\tilde{v}_2(X))$. These corrections of scores for the effects of nuisance parameters will show up again in the derivation of properties of the likelihood ratio test.

For its practical value, we indicate how the variance of the test statistic can be calculated when there are other auxiliary parameters in the problem, say $\theta_1, \ldots, \theta_m$. If we let $u_1, \ldots, u_m$ be the score functions for the auxiliary parameters, then one constructs the $m+2$ by $m+2$ covariance matrix i^* of the extended set of scores $u_1, \ldots, u_m, v_1, v_2$. Other than the last row and column, this is the Fisher information matrix i for the unicomponent problem. We write it in the partitioned form

$$i^* = \begin{bmatrix} i & a \\ a' & b \end{bmatrix}.$$

The formula for the asymptotic variance of $v_2(\hat{\theta}_1, \ldots, \hat{\theta}_m, \hat{\phi}; x)$ is then $b - a' i^{-1} a$.

4.1.3. *Dispersion test optimality.* A remarkable and important feature of the $C(\alpha)$ test for heterogeneity is that the test statistic does not depend on the alternative distribution G that was used and so it suggests that the statistic has power over a wide range of nearby alternatives. (If we think about the statistic in terms of the gradient, this becomes quite plausible.) Indeed, the statistic is often derived through other means than presented here, such as constructing the locally most powerful test for degeneracy in a conjugate family of latent densities.

However, a curiosity exists in the literature. Moran (1973) found that he could prove that this test, in the case of the test for Poisson overdispersion, was asymptotically best in the sense of best power under local alternatives, *only* if he assumed in addition that

$$m_3(G) := \int s^3 \, dG(s) = 0.$$

It appears that this condition is not strictly necessary, as we argue now briefly.

If we carry out a Taylor series expansion of the sample log likelihood about $b = 0$, we obtain the formula

$$\begin{aligned} \sum \ln\left[\frac{f(x_i; a, b)}{f(x_i; a, 0)}\right] &= b \cdot 0 + \frac{b^2}{2!} \sum v_2(a; x_i) + \frac{b^3}{3!} m_3(G) \sum v_3(a; x_i) \\ &\quad + \frac{b^4}{4!} \left[m_4(G) \sum v_4(a; x_i) - 3 \sum v_2(a; x_i)^2 \right] \\ &\quad + \text{remainder}. \end{aligned} \tag{4.4}$$

Here we have used the higher order versions of the dispersion score:

$$v_k(a; x) := \frac{f^{(k)}(x; a)}{f(x; a)}.$$

Noting that the functions $v_k(a; X)$ all have mean *zero* under the null hypothesis, we find that the log likelihood ratios *converge to normality* if we set

$b = b_n = cn^{-1/4}$. With such a scaling the quadratic term is asymptotically normal, the cubic term converges to *zero, for any finite* $m_3(G)$, and for finite $m_4(G)$ the quartic term converges to the constant,

$$-\frac{3c^4 E[v_2(a;X)^2]}{4!}.$$

From this point it appears that by taking local alternatives that approach the null at the unusual rate of $n^{-1/4}$ in the parameter b, one can derive the necessary asymptotic optimality, as well as local power.

The key here seems to be that instead of following the above approach, Moran considered the $n^{-1/2}$ convergence of the parameter b^2. We note that if we use the scale parameter b, then the parametric family of likelihoods generated by $\Phi = a+b\Theta$ is in fact well defined for b both positive and negative, and generates a smooth two-sided alternative to the null hypothesis. If Θ has a symmetric distribution G, then $-b$ generates the same distribution as b, but otherwise not. If not, however, this means that the family of likelihoods is not symmetric as a function of b, and so is not a function of b^2 and so cannot have a Taylor expansion in the variable b^2.

Thus, although the $\sqrt{n}$ convergence of b^2 might make it seem to be the appropriate parameterization, switching to the more slowly converging scale parameter b makes the expansions work properly, because then they are valid for b both positive and negative.

The fact that one has asymptotic power, in the b scale, only at distances of order $n^{-1/4}$, has importance in the general nonparametric theory as well. Chen (1993) used a derivation like that above, with G being a two point distribution, to show that the optimal rate of convergence of a consistent G estimator to an unknown discrete latent distribution is $n^{-1/4}$ if the number of components is not known or has been overspecified.

4.1.4. *Auxiliary parameters.* As a final note, we make the observation that further problems may arise in certain mixture models with auxiliary parameters. To give a simple example, if we consider the normal mixture model $N(G, \sigma^2)$, it is easily seen that the dispersion score v_2 is *equivalent* to the nuisance parameter score for σ^2. Since the nuisance parameter scores must be regressed out of the locally most powerful test statistic, it follows that the locally most powerful tests for heterogeneity from (4.4) are no longer independent of the form of G. If $m_3(G)$ is *not zero*, then the locally most powerful test is based on the corrected version of the "skewness" score v_3. If $m_3(G)$ is *zero*, then the best test depends on the "kurtosis" score v_4. In either case, the rate of convergence of detectable alternatives to the null must be slower yet than $n^{-1/4}$.

Intuitively, this arises because of an identifiability issue. If the latent distribution G is $N(v, \tau^2)$, then X has a distribution that can be represented as either a mixture $N(G, \sigma^2)$ or a different normal $N(v, \sigma^2 + \tau^2)$. Thus any hope of detecting a mixture alternative will depend on the degree to which the G differs from the normal, as evidenced through its cumulants.

4.2. LRT for number of components. In this section we will give some background on the likelihood ratio test (LRT) for the number of components in a mixture model. The last section of this chapter will give some new results regarding the limiting distributions involved, both for the test of one versus two components, and the test of one component versus the nonparametric model. However, it is important to point out that from a practical point of view, the $C(\alpha)$ test is much simpler to calculate and has a much nicer distribution theory.

4.2.1. *The testing problem.* The nature of the limiting distribution for this likelihood ratio test is a long-standing mystery. If we consider just the simple model with no auxiliary parameters and where ϕ is unidimensional, then the usual distribution theory suggests that the likelihood ratio test has a chi-squared distribution with 2 degrees of freedom, corresponding to one free parameter under the null hypothesis H of one component, with density $f(x;\phi)$, and three parameters under the alternative A_2, with density

$$\pi f(x;\phi_1) + (1-\pi) f(x;\phi_2).$$

However, it is known that the usual regularity conditions are not satisfied.

To consider the difficulties, we consider the parameter space for the two-component model, where we can restrict attention to

$$\pi \in [0,1] \quad \text{and} \quad \phi_1 \leq \phi_2.$$

In this setting we can describe a single element of the null hypothesis, say unicomponent with parameter ϕ_0, with many elements of the alternative parameter space. We find three lines on the boundary of the parameter space all give the same null distribution:

- The line for $\phi_1 = \phi_2 = \phi_0$, $\pi =$ anything.
- The line where $\pi = 0$, $\phi_2 = \phi_0$ and $\phi_1 =$ anything.
- The line where $\pi = 1$, $\phi_1 = \phi_0$ and $\phi_2 =$ anything.

In Figure 4.1 we plot the alternative parameter values that correspond to a single null distribution, unicomponent with parameter ϕ_0.

The union of all such lines, over all values of ϕ_0, makes up the null hypothesis and so it corresponds to three entire boundary surfaces of the alternative parameter space. It is clear from this description that there are many points in the alternative that seem to be close to any one distribution in the null hypothesis.

The first fundamental difficulty in establishing the asymptotic structure of the likelihood ratio test lies in determining what happens to the score functions in the alternative hypothesis as we approach the null. Because one can approach the null from so many directions, we will find that the set of limit functions at the null hypothesis is an *infinite*-dimensional score space, even though there are only three scores in the alternative hypothesis, corresponding to three parameters. In turn, this structure implies that the problem generates what we will call a type III likelihood ratio test, a rather awkward beast

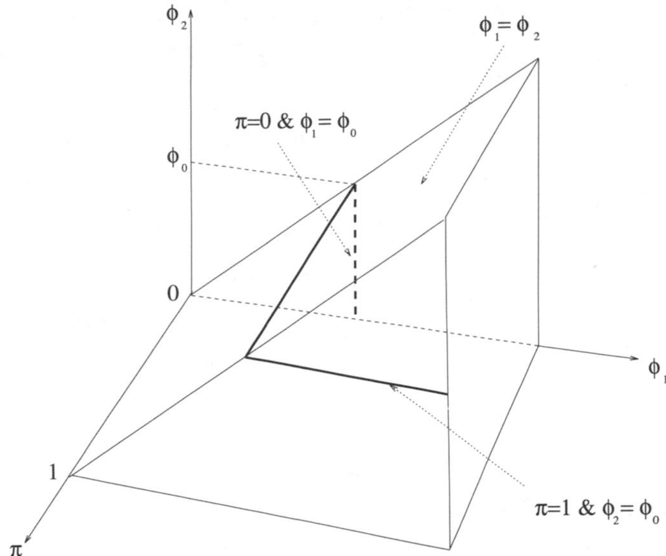

FIG. 4.1. *The parameter space for the two-component mixture.*

at best. (Type I is the standard testing problem with limiting chi-squared distributions; type II generates the chi-bar-squared distributions.)

4.2.2. *Historical perspective.* For background on the problem of likelihood ratio testing in mixture models, the books by Titterington, Smith and Makov (1985) and McLachlan and Basford (1988) provide extensive discussion. Over the years considerable simulation work has been done; see Böhning, Dietz, Schaub, Schlattman and Lindsay (1994) for a variety of exponential family simulations. In all, the limiting distributions are not completely apparent from such studies, but the normal theory distribution, in the unknown variance case, has appeared tantalizingly like the chi-squared with 2 degrees of freedom, and the Böhning study shows that one parameter exponential families appear to have tails very much like mixtures of chi-squares with differing degrees of freedom; these results are in accordance with the distributions to be derived in the next section.

Exact theoretical results have been obtained in a number of special cases. For example, if we have two known components, with unknown π only, and the null hypothesis is $\pi = 0$, then it will be shown in the next section that the limiting distribution is

$$0.5\chi_0^2 + 0.5\chi_1^2.$$

Here the symbol χ_0^2, the "chi-squared with 0 degrees of freedom," is a degenerate distribution with all its mass at 0. Such a mixture of chi-squared distributions of different degrees of freedom is called a *chi-bar-squared* distribution. Another interesting set of special cases was considered by Goffinet,

Loisel and Laurent (1992), who were interested in normal model hypotheses when the weight π was known a priori. The limiting distributions were again of the chi-bar-squared type.

The difficulties with the irregular parameter space in the general problem led Aitkin and Rubin (1985) to attempt to make the problem regular by using a prior distribution on π. However, Quinn, McLachlan and Hjort (1987) showed that even with this approach the usual regularity conditions were violated.

Ghosh and Sen (1985) developed a theory for models that satisfy a rather severe identifiability constraint. The most relevant and important insight into the general problem can be found in Hartigan (1985), who established that in the normal model, if one of the ϕ's is known and if σ^2 is known, then the likelihood ratio test statistic has no interesting limiting distribution, but rather goes to *infinity* with probability 1.

4.2.3. *Initial observations.* For the moment we assume the model has no auxiliary parameters. We first make some observations about the probability that the likelihood ratio statistic equals zero, corresponding to having a χ_0^2 component in the limiting distribution. We let $\hat{\phi}$ be the maximum likelihood estimator under the null hypothesis and make the following claim:

$$\text{LRT} = 0 \quad \Leftrightarrow \quad D_1(\phi) \leq 0 \qquad \forall \phi. \tag{4.5}$$

Check this, using the fact that the gradient function D_1 satisfies the right-hand inequality if and only if $\Delta_{\hat{\phi}}$ is the nonparametric maximum likelihood estimator of Q. Further, it follows from the asymptotic normality of the Neyman–Scott dispersion test statistic and relationship of the dispersion test to the gradient inequality as in (4.3) that the gradient inequality is violated in a neighborhood of $\hat{\phi}$ with an asymptotic null probability of at least 0.5. It follows that there is asymptotically *at least* probability 0.5 that the LRT is greater than zero.

The gradient equation (4.5) makes it straightforward to do simulations of the probability that the likelihood ratio statistic takes on the value 0. However, if the gradient inequality fails and we are testing one component versus two, then there are some fundamental difficulties with simulation studies due to the nonuniqueness of the solutions to the likelihood equations. Two simulation studies that use different algorithms, different starting values or different convergence criteria are studying different statistics. Unfortunately, we know very little about how much difference it might make.

From a practical point of view, one would like the simulation study to exactly mimic the procedure one uses on a data set, so that the results are appropriate to the estimator actually used. An approach used by Furman and Lindsay (1994b) was to use the unique and easy-to-calculate moment estimators as initial values, then iterate 100 times with the EM algorithm. Since this process is easy to replicate, it gives a fast and useful way to construct critical values by simulation.

In the face of this, it seems even more important to understand the asymptotic structure of the mixture problem better, especially the nature of the multimodality problem.

4.3. Asymptotic multinomial geometry. We now offer a side trip into a particular geometric formulation of asymptotics in the multinomial model, designed to be background for the analysis of likelihood ratio testing in the mixture model. The reader may find that his needs are suited by skimming this section and considering just the implications presented in the next section of this chapter. The results found in this section are not new, but the geometric formulation of the asymptotics is presented with more completeness than is to be found elsewhere, as far as I know.

We return to a general discrete density $f(t;\eta)$ for $t \in \{0, 1, \ldots, T\}$, written in vector form as $\mathbf{f}_\eta$. We suppose that we have an i.i.d. sample of size n and that $\mathbf{d}$ is the vector of sample proportions. We suppose for the probability calculations that $\eta = \eta_0$ is the true parameter value, with corresponding vector $\mathbf{f}_0 := \mathbf{f}_{\eta_0}$. The asymptotic result used to drive the analysis is the fact that as $n \to \infty$, we have

$$n^{1/2}(\mathbf{d} - \mathbf{f}_0) \to \mathrm{MVN}(\mathbf{0}, \mathbf{V}).$$

Here the matrix $\mathbf{V} = \mathrm{diag}(\mathbf{f}_0) - \mathbf{f}_0\mathbf{f}_0'$. The covariance matrix is rank deficient since the probability vectors are constrained to lie in the T-dimensional probability simplex, now denoted $\mathbf{P}$.

4.3.1. *The dagger simplex.* To set up the geometry, it is useful to perform a transformation of the vectors in the simplex that will give our calculations a natural probabilistic interpretation. The transformed space arises from constructing ratios of densities with respect to the true density f_0. For any function $g(t)$ on the sample space, we define the *dagger operation* to be

$$g^\dagger(t) := \frac{g(t)}{f_0(t)} - 1.$$

Note that the inverse of the dagger operation is $g(t) = f_0(t)[g^\dagger(t) + 1]$ and that we are necessarily assuming that the true density f_0 is strictly positive on the sample space.

If we apply the dagger operation to all elements of the simplex, $\mathbf{p} \to \mathbf{p}^\dagger$, we obtain a transformed simplex $\mathbf{P} \to \mathbf{P}^\dagger$. The dagger simplex is the convex hull of the extreme points $\mathbf{e}_k^\dagger$. Most importantly for our calculations, the dagger simplex lies in the hyperplane of vectors orthogonal to $\mathbf{f}_0$:

$$\mathbf{P}^\dagger \subset \mathscr{M}_0 = \{\mathbf{v}\colon\ \mathbf{v} \cdot \mathbf{f}_0 = 0\}.$$

Note also that the stochastic interpretation of the equation $\mathbf{v} \cdot \mathbf{f}_0 = 0$ is that $v(X)$, if $X \sim f_0$, is a mean zero variable: $\mathbf{v} \cdot \mathbf{f}_0 = \sum v(t) f_0(t)$. Thus $\mathscr{M}_0$ is the linear space of mean zero variables under the null model. We note moreover that the dagger transformation took the true density and placed it at the origin: $\mathbf{f}_0^\dagger = \mathbf{0}$. It is important to remember that the origin plays the role of a selected element of the null hypothesis.

On this space we will use the E_0 inner product defined by expectations under E_0 :

$$\langle \mathbf{g}^\dagger, \mathbf{h}^\dagger \rangle = E_0[g^\dagger(X)h^\dagger(X)] = \sum f_0(t)g^\dagger(t)h^\dagger(t).$$

In this space, the distance between two vectors is then

$$\|\mathbf{g}^\dagger - \mathbf{h}^\dagger\|^2 = E_0[g^\dagger(X) - h^\dagger(X)]^2 = \sum \frac{[g(t) - h(t)]^2}{f_0(t)},$$

a form of chi-squared distance on the undaggered functions.

If we let $\mathbf{F}$ be the surface of model densities in the simplex,

$$\mathbf{F} = \{\mathbf{f}_\eta\colon\ \eta \in \text{parameter space}\},$$

then there is a corresponding model surface $\mathbf{F}^\dagger$ in the dagger simplex, and it contains the origin (the selected null model). If η is p-dimensional, we let $\mathbf{s}_1, \mathbf{s}_2, \ldots, \mathbf{s}_p$ be the score function vectors at $\eta = \eta_0$; that is, the vectors with tth component defined by

$$s_j(t,\eta) = \frac{\partial \ln f(t;\eta)}{\partial \eta_j}.$$

The score functions give us a way to approximate the space of models $\mathbf{F}^\dagger$ with a linear manifold on which it is easier to do asymptotics. If we pick any *direction* $\mathbf{h}$ in the parameter space, then a first order Taylor expansion gives

$$\mathbf{f}^\dagger_{\eta_0+\mathbf{h}} = (\mathbf{f}^\dagger_{\eta_0+\mathbf{h}} - \mathbf{0}) \approx (\mathbf{s}_1, \ldots, \mathbf{s}_p)\mathbf{h} = \sum h_i \mathbf{s}_i. \tag{4.6}$$

That is, in this space, elements of the model surface $\mathbf{F}^\dagger$ can be approximated in a neighborhood of the null model by linear combinations of the score vectors, the *score tangent space*

$$\mathbf{S} = \left\{\sum h_i \mathbf{s}_i\colon\ h_i \in \mathbf{R}\right\}. \tag{4.7}$$

This approximation is pictured in Figure 4.2. Note also that the E_0 squared length of a score vector is just the corresponding *Fisher information* $E_0[s_j(X)]^2$. Thus the lengths in this space have a natural statistical interpretation.

Further, and this is quite important for our investigation, if there are *model constraints* on η that limit the directions $\mathbf{h}$ one can move in the parameter space away from η_0, the approximating tangent surface $\mathbf{S}$ will have corresponding constraints on the coefficients h_i. For example, if $\eta_0 = (0, \ldots, 0)'$ and if within the model $\mathbf{F}$ some of the η_i are restricted to be nonnegative, then, from (4.6), the approximating surface (4.7) should have the same restriction on the h_i.

4.3.2. *Maximum likelihood and projections.* The next goal is to reexpress the asymptotics of maximum likelihood estimation in the language of this geometry via studying the properties of the *projections* of the sample proportions onto the tangent score space, which might be called *data-to-model projections*, or more simply, *data projections*.

To do so, we consider the geometric relationship between the daggered data vector $\mathbf{d}^\dagger$ and maximum likelihood estimation. First we note that the E_0 squared distance between $\mathbf{d}^\dagger$ and true density $\mathbf{f}_0^\dagger = \mathbf{0}$ is

$$\sum \frac{[d(t) - f_0(t)]^2}{f_0(t)},$$

which is n^{-1} times the Pearson chi-squared statistic for testing

$$H\colon\ f(t) = f_0(t)$$

against a general multinomial alternative. It therefore has a limiting chi-squared distribution with T degrees of freedom.

Let $\mathscr{P}_{\mathbf{S}}\mathbf{v}$, for $\mathbf{v} \in \mathbf{P}^\dagger$, be the E_0 *projection* of $\mathbf{v}$ onto the linear space $\mathbf{S}$ of the scores. That is,

$$\mathscr{P}_{\mathbf{S}}\mathbf{v} = \sum b_i \mathbf{s}_i,$$

where the coefficients b_i are chosen to minimize the E_0 distance

$$\left\| \mathbf{v} - \sum b_i \mathbf{s}_i \right\|.$$

Let $\hat{\eta}$ be the maximum likelihood estimator of the parameter η, with $\hat{\mathbf{f}} := \mathbf{f}_{\hat{\eta}}$ thereby being the maximum likelihood estimator of the density vector. Our *claim* is that the data projection $\mathscr{P}_{\mathbf{S}}\mathbf{d}^\dagger$ is asymptotically equivalent to the MLE $\hat{\mathbf{f}}^\dagger$ in the sense that

$$n\|\hat{\mathbf{f}}^\dagger - \mathscr{P}_{\mathbf{S}}\mathbf{d}^\dagger\|^2 \to 0 \quad \text{in probability.}$$

This is pictured in Figure 4.2.

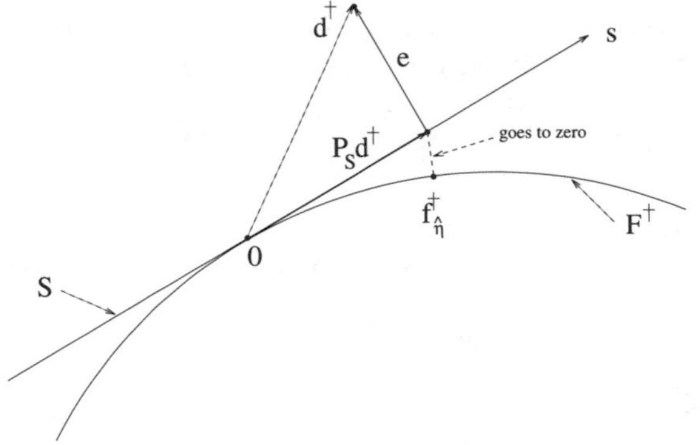

FIG. 4.2. *The score tangent space approximates the daggered model.*

For simplicity's sake, we consider the proof of the claim in the $(p = 1)$-dimensional case. First, a Taylor expansion of $\hat{\mathbf{f}}^\dagger$ in η about η_0 gives us

$$\hat{\mathbf{f}}^\dagger = \mathbf{0} + (\hat{\eta} - \eta_0)\mathbf{s} + O_p(n^{-1}).$$

[We note, as an important aside for novices in asymptotics, that this implies that "all the action" is taking place near the origin because $(\hat{\eta} - \eta_0) = O_p(n^{-1/2})$. As $n \to \infty$, provided the null model is true, we can count on the statistical objects of interest becoming closer and closer to the origin, which is where the tangent score approximation works best.]

Second, we can explicitly calculate the projection of $\mathbf{d}^\dagger$ on the score space to be

$$\mathscr{P}_{\mathbf{S}}\mathbf{d}^\dagger = \hat{b}\mathbf{s} \quad \text{where } \hat{b} = \frac{\langle \mathbf{s}, \mathbf{d}^\dagger \rangle}{\|\mathbf{s}\|^2} = \sum d(t)s(t;\eta_0)i^{-1}, \tag{4.8}$$

where i is the Fisher information about η at η_0. [*Exercise.*] However, examination of $\hat{b}$ in the last equation shows it is equal to the first order influence function expansion of the maximum likelihood functional $\hat{\eta}$ about η_0, so that $\hat{\eta} - \eta_0 = \hat{b} + O_p(n^{-1})$. Hence

$$n\|\hat{\mathbf{f}}^\dagger - \mathscr{P}_{\mathbf{S}}\mathbf{d}^\dagger\|^2 = n\|(\hat{\eta} - \eta_0 - \hat{b})\mathbf{s} + O_p(n^{-1})\|^2 = O_p(n^{-1}).$$

This establishes our claim.

Thus finding the maximum likelihood estimator of the density is, to the appropriate statistical order, equivalent to finding a projection of the daggered data onto the tangent score space.

Finally, we establish a simple asymptotic distribution theory for data projections. Note that taking inner products with the daggered data vector gives us sample averages:

$$\langle \mathbf{v}, \mathbf{d}^\dagger \rangle = \sum_t f_0(t)v(t)\left(\frac{d(t)}{f_0(t)} - 1\right) = n^{-1}\sum_i v(X_i),$$

provided $v(t) \in \mathscr{M}_0$. Thus for elements $\mathbf{v} \in \mathscr{M}_0$ there exists a very simple rule for calculation of limiting distributions for pairs (and vectors) of inner products with the daggered data vector. It is a multivariate normal distribution, with E_0 inner products determining the covariance matrix:

$$n^{1/2}\begin{bmatrix} \langle \mathbf{v}, \mathbf{d}^\dagger \rangle \\ \langle \mathbf{w}, \mathbf{d}^\dagger \rangle \end{bmatrix} \to N\left[\begin{pmatrix} 0 \\ 0 \end{pmatrix}, \begin{pmatrix} \|\mathbf{v}\|^2 & \langle \mathbf{w}, \mathbf{v} \rangle \\ \langle \mathbf{w}, \mathbf{v} \rangle & \|\mathbf{w}\|^2 \end{pmatrix}\right]. \tag{4.9}$$

This is a simple *exercise* for the reader.

This leads to an elegant expression for the limiting distribution for a data projection. Suppose we have a linear space $\mathbf{T}$ with an *orthonormal basis* (E_0 inner product) $\mathbf{t}_1, \ldots, \mathbf{t}_k$, so that $\|\mathbf{t}_j\| = 1$ and $\langle \mathbf{t}_i, \mathbf{t}_j \rangle = 0$. The use of such a basis is desirable, because then the projection of $\mathbf{d}^\dagger$ onto $\mathbf{T}$ can be represented very simply as the vector

$$\langle \mathbf{t}_1, \mathbf{d}^\dagger \rangle \mathbf{t}_1 + \cdots + \langle \mathbf{t}_k, \mathbf{d}^\dagger \rangle \mathbf{t}_k.$$

This leads to a *fundamental result* that will simplify many of the calculations that we will undertake in the following. From (4.9), the asymptotic distribution of the data projection is therefore

$$n^{1/2}\mathscr{P}_{\mathbf{T}}(\mathbf{d}^{\dagger}) \to Z_1\mathbf{t}_1 + \cdots + Z_k\mathbf{t}_k, \tag{4.10}$$

where $(Z_1, \ldots Z_k)$ are i.i.d. standard normal variates.

4.3.3. *Type I likelihood ratio testing.* We now turn to the geometric interpretation of the likelihood ratio tests. Our claim is that the likelihood ratio tests for model hypotheses are asymptotically equivalent to the lengths of certain E_0 projections in the tangent score space.

First, we consider the simple versus composite hypotheses $\eta = \eta_0$ versus $\neq$. Here we assume knowledge from standard asymptotic results available elsewhere that the quadratic form of the score test statistic is asymptotically equivalent to the likelihood ratio test statistic. It is easily checked that the normalized quadratic form score test statistic

$$n\left[\sum d(t)\mathbf{s}(\eta_0, t)\right]i^{-1}\left[\sum d(t)\mathbf{s}(\eta_0, t)\right]$$

for these hypotheses is $n\|\mathscr{P}_{\mathbf{S}}(\mathbf{d}^{\dagger})\,\|^2$. [*Exercise*: See (4.8).] Moreover, we note that, in general, n times the squared length of the projection of $\mathbf{d}^{\dagger}$ onto a linear space such as $\mathbf{S}$ will result in a variate that has a limiting chi-squared distribution, with degrees of freedom equal to the dimension of $\mathbf{S}$, which is the same as the number of parameters in the model. [*Exercise*: Work this out using (4.10).]

The next level of difficulty is to incorporate nuisance parameters into the testing problem. We partition $\eta = (\theta, \gamma)$ and consider the hypotheses $\theta = \theta_0$ versus $\neq$. We now have two sets of score functions corresponding to the two sets of parameters. We will denote the θ scores by $\mathbf{u}$'s and the γ scores by $\mathbf{v}$'s. We then decompose the score tangent space $\mathbf{S}$ into two parts: the *nuisance score space* $\mathbf{V}$, generated by linear combinations of the $\mathbf{v}_j$, and the *corrected score space* $\tilde{\mathbf{U}}$, which will be the orthogonal complement of $\mathbf{V}$ within $\mathbf{S}$. Note that if we construct *corrected scores* via

$$\tilde{\mathbf{u}}_k := \mathbf{u}_k - \mathscr{P}_{\mathbf{V}}(\mathbf{u}_k),$$

then $\tilde{\mathbf{U}}$ is generated by linear combinations of the $\tilde{\mathbf{u}}_k$. (These are also known as the *efficient* scores.)

In this case, the likelihood ratio statistic for the composite hypotheses $\theta = \theta_0$ versus $\neq$ corresponds asymptotically to the difference in two score test statistics. That is, we can write

$$\begin{aligned}2[\ln f(x;\hat{\theta},\hat{\gamma}) &- \ln f(x;\theta_0,\hat{\gamma}_{\theta_0})]\\ &= 2[\ln f(x;\hat{\theta},\hat{\gamma}) - \ln f(x;\theta_0,\gamma_0)]\\ &\quad - 2[\ln f(x;\theta_0,\hat{\gamma}_{\theta_o}) - \ln f(x;\theta_0,\gamma_0)],\end{aligned}$$

which is the difference between the test statistics for two simple versus composite hypotheses and so is asymptotically equivalent to

$$n\|\mathscr{P}_{\mathbf{S}}(\mathbf{d}^{\dagger})\|^2 - n\|\mathscr{P}_{\mathbf{V}}(\mathbf{d}^{\dagger})\|^2 = n\|\mathscr{P}_{\tilde{\mathbf{U}}}(\mathbf{d}^{\dagger})\|^2.$$

The last equality above derives from orthogonality of $\mathbf{V}$ and $\tilde{\mathbf{U}}$ and the fact that they span $\mathbf{S}$. The test statistic has, therefore, from the above remarks, an asymptotic chi-squared distribution with degrees of freedom equal to the dimension of θ. A testing situation in which the above standard likelihood ratio asymptotic theory applies will be called a *type I* problem.

4.4. The type II likelihood ratio problem. The next step in our analysis is to take account of *restrictions* on the parameter space that will alter the preceding distribution theory. This section increases the level of difficulty to the next degree. Although it is not sufficient to carry us all the way through to the solution of the mixture likelihood ratio solution, it is essential background.

4.4.1. *Parameter constraints.* Inasmuch as this section is designed to lead to the next, we will simplify the general context somewhat. In the general context, there are *focal* parameters θ, parameters of interest that are restricted by the null hypothesis and *nuisance* parameters γ that are not specified by the hypotheses. In general, both the focal and nuisance parameters will be divisible into two types: those constrained by the model and those not. (This description is a bit vague, but we hope that further considerations will clarify it.) In order to be true to our objective—the mixture model LRT—and keep things as simple as possible, we will reduce to the case where the focal parameters *all* have constraints put upon them in the neighborhood of the null hypothesis, but that *none* of the nuisance parameters do. To further simplify, we will assume that the null hypothesis is $\boldsymbol{\theta} = \mathbf{0}$ and that the alternative is $\boldsymbol{\theta} \geq \mathbf{0}$.

As they will be useful to us the next section, we also will follow throughout this section the following two examples.

EXAMPLE 11. We have a single focal parameter θ, with the restriction that $\theta \geq 0$. The null hypothesis is $\theta = 0$ and the alternative is $\theta \geq 0$. All nuisance parameters are unconstrained in the neighborhood of the true model (θ_0, γ_0).

EXAMPLE 12. In addition to an arbitrary set of nuisance parameters, we have two focal parameters (θ_1, θ_2), with the constraint under the alternative:

$$\theta_1 \geq 0, \qquad \theta_2 \geq 0.$$

4.4.2. *Convex cones.* As we noticed earlier, under a set of restrictions the model surface $\mathbf{F}^{\dagger}$ is no longer approximated by $\mathbf{S}$, the entire linear space generated by the scores $\mathbf{s}_i$, but rather by a restricted set, in our case the *score tangent cone:*

$$\mathbf{S}^* = \left\{\sum a_i\mathbf{u}_i + \sum b_j\mathbf{v}_j\colon\ a_i \geq 0,\ b_j \in \mathscr{R}\right\}.$$

It is useful to note that we can also write $\mathbf{S}^*$ in terms of the corrected focal scores:

$$\mathbf{S}^* = \left\{ \sum a_i \tilde{\mathbf{u}}_i + \sum b_j \mathbf{v}_j \colon a_i \geq 0,\ b_j \in \mathscr{R} \right\}.$$

At this point we need some further terminology to describe the kind of set we are dealing with. First, a set of vectors $\mathbf{W}$ is a *cone* if it contains all *rays* through points in $\mathbf{W}$: that is, if $\mathbf{w} \in \mathbf{W}$, then $c\mathbf{w} \in \mathbf{W}$ for all $c > 0$. The set $\mathbf{W}$ is a *convex cone* if it is a cone that is also convex. Check that the restricted tangent space $\mathbf{S}^*$ is a convex cone.

We will call $\mathbf{W}$ the *positive cone* generated by the vectors $\{\mathbf{w}_1, \ldots, \mathbf{w}_k\}$ if

$$\mathbf{W} = \left\{ \sum a_i \mathbf{w}_i \colon a_i \geq 0 \right\}.$$

A positive cone is clearly a convex cone.

Although convex cones do not have all the desirable features of a linear space, if they are *closed*, they do share with a linear space the uniqueness of projections. That is, given any vector $\mathbf{v}$ and a closed convex cone $\mathbf{W}$, there is a unique vector $\mathbf{w} \in \mathbf{W}$, *the projection of* $\mathbf{v}$ *onto* $\mathbf{W}$, that minimizes $\|\mathbf{v} - \mathbf{w}\|$. We will study some further properties of these projections later.

Chernoff (1954) proved that the limiting distribution theory of the likelihood ratio test can be generated by doing data projections onto the score tangent cones, provided that these cones approximate the model surface. In our example, this means that the likelihood ratio statistic is asymptotically equivalent to

$$n\|\mathscr{P}_{\mathbf{S}^*}(\mathbf{d}^\dagger)\|^2 - n\|\mathscr{P}_{\mathbf{V}}(\mathbf{d}^\dagger)\|^2,$$

where $\mathbf{V}$ is the tangent space of the nuisance parameters. Check that this is also the squared length of the data projection $\mathscr{P}_{\tilde{\mathbf{U}}^*}(\mathbf{d}^\dagger)$, where

$$\tilde{\mathbf{U}}^* := \left\{ \sum a_i \tilde{\mathbf{u}}_\mathbf{i} \colon a_i \geq 0 \right\}$$

is the positive cone generated by the corrected score functions. (See Figure 4.3.) That is, we may without loss of generality restrict our attention to the linear space $\tilde{\mathbf{U}}$ that contains $\tilde{\mathbf{U}}^*$ and ignore the orthogonal directions corresponding to the space $\mathbf{V}$.

4.4.3. *The z-coordinate system.* It is useful to define a new coordinate system for the linear space $\tilde{\mathbf{U}}$. If we do so appropriately, then we can reduce the general problem to a standardized form involving projections of i.i.d. standard normal random variables, so that we can directly deduce the limiting distributions of the projections. Returning to the original coordinate system then will give us the implications for a general model.

We taken any E_0 orthonormal basis for $\tilde{\mathbf{U}}$, say $\mathbf{b}_1, \ldots, \mathbf{b}_d$, and represent points $\mathbf{w}$ in this space by their orthonormal coordinates $\mathbf{z} = \mathbf{z}(\mathbf{w})$ defined by

$$z_1 = \langle \mathbf{b}_1, \mathbf{w} \rangle, \ldots, z_d = \langle \mathbf{b}_d, \mathbf{w} \rangle.$$

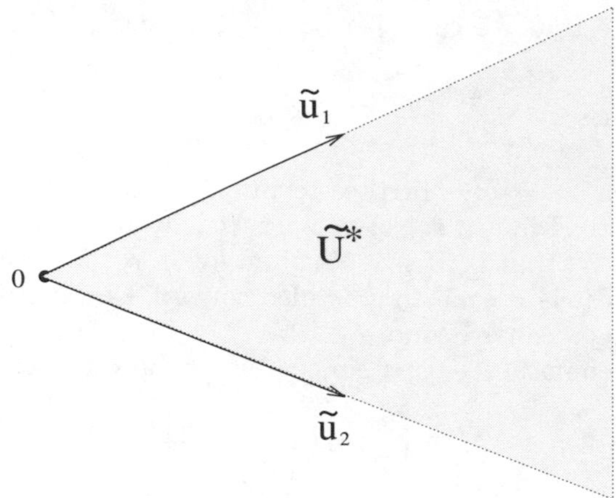

FIG. 4.3. *The positive cone generated by two corrected score functions.*

[Of course, there exists an appropriate $d \times (T+1)$ matrix $\mathbf{B}$ such that $\mathbf{Bw} = \mathbf{z}$.] We will let the z-coordinate representation of the normalized data projection

$$n^{1/2}\mathscr{P}_{\tilde{\mathbf{U}}}\mathbf{d}^{\dagger} = \mathscr{P}_{\tilde{\mathbf{U}}}(n^{1/2}\mathbf{d}^{\dagger})$$

be denoted by $\mathbf{Z}$. Recall from our original considerations (4.10) that $\mathbf{Z}$ is asymptotically standard normal.

We have thus replaced our original data with a set of standard normal variables. We next need to understand how the restrictions in the original score space show up in this transformed version and how the original E_0 geometry is transformed.

The first simplification comes because when working with the z-coordinates we replace the E_0 geometry with the ordinary Euclidean inner product and distance. This arises because the E_0 inner product between any two points in $\tilde{\mathbf{U}}$ equals the ordinary Euclidean inner product for their $\mathbf{z}$-coordinate representations:

$$\begin{aligned}\left\langle \sum z_{i1}\mathbf{b}_i, \sum z_{i2}\mathbf{b}_i \right\rangle &= E_0\left[\left(\sum z_{i1}b_i(X)\right)\left(\sum z_{i2}b_i(X)\right)\right] \\ &= \sum z_{i1}z_{i2} \\ &= \mathbf{z}_1 \cdot \mathbf{z}_2.\end{aligned}$$

In this new coordinate system, we have a transformed version $\mathbf{z}(\tilde{\mathbf{U}}^*)$ of the original cone $\tilde{\mathbf{U}}^*$, corresponding to the set of coordinates $\mathbf{z}$ of all the points in $\tilde{\mathbf{U}}^*$. As an *exercise*, show that it is generated as the positive cone of the normalized extremal vectors

$$\mathbf{p}_i := \mathbf{z}(\tilde{\mathbf{u}}_i)/\|\mathbf{z}(\tilde{\mathbf{u}}_i)\|.$$

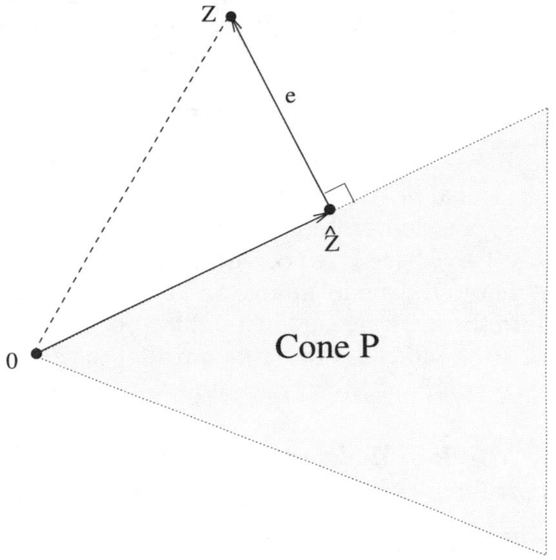

FIG. 4.4. *The projection of* $\mathbf{z}$ *onto the cone* $\mathbf{P}$.

We will call this cone in z-space

$$\mathbf{P} = \left\{ \sum a_i \mathbf{p_i} \colon \ a_i \geq 0 \right\}$$

the *primal cone*.

Suppose we wish to find the z coordinates of the E_0 projection of $\mathbf{d}^\dagger$ onto $\tilde{\mathbf{U}}^*$ so that we can determine the limiting distribution of our data projection. The projection can be carried out in two steps, by first projecting onto $\tilde{\mathbf{U}}$, then onto $\tilde{\mathbf{U}}^*$. Because of the equivalence above for inner products, and therefore distances, we can *instead* work in the z-coordinate space and use Euclidean distances. That is, we take the z-coordinates of $\mathscr{P}_{\tilde{\mathbf{U}}}\mathbf{d}^\dagger$ and do an ordinary Euclidean projection of them onto $\mathbf{P}$.

Thus we have transformed our asymptotic problem into the problem of determining the distribution of the squared length of the projection of a vector of *standard normal variables* $\mathbf{Z}$ onto a convex cone $\mathbf{P}$. The situation is pictured in Figure 4.4, where the projection is denoted $\hat{\mathbf{Z}}$.

4.4.4. *Projections onto convex cones.* We now study the properties of projections onto convex cones because then we can determine the limiting distributions of the data projections. A useful concept, akin to the notion of the orthogonal subspace, is the idea of the dual convex cone.

The *dual (or polar) cone* $\mathbf{P}^o$ to a convex cone $\mathbf{P}$ is defined to be the set of all vectors that are negatively correlated with all the vectors in $\mathbf{P}$, namely,

$$\mathbf{P}^o = \{\mathbf{y} \colon \ \mathbf{y} \cdot \mathbf{m} \leq 0 \text{ for all } \mathbf{m} \in \mathbf{P}\}.$$

If the cone $\mathbf{P}$ is closed and convex, then $\mathbf{P}^o$ is closed and convex and $(\mathbf{P}^o)^o = \mathbf{P}$. To simplify the notation, let

$$\hat{\mathbf{z}} := \mathscr{P}_{\mathbf{P}}\mathbf{z} \quad \text{and} \quad \mathbf{e} := \mathscr{P}_{\mathbf{P}^o}\mathbf{z}$$

be the *primal* and *dual projections* of $\mathbf{z}$. The terminology *fitted value vector* and *residual vector* would also be appropriate, as a reminder of the similarity of this problem to the regression problem, in which the vector of observations $\mathbf{y}$ is decomposed into $\hat{\mathbf{y}} + \mathbf{e}$, where $\hat{\mathbf{y}}$ is the vector of fitted values, determined by linear projection of $\mathbf{y}$ onto the model space, and $\mathbf{e}$ are the residuals, determined by the projection onto the subspace orthogonal to the model. The second part of the following proposition indicates that the parallel to the regression problem extends to the orthogonality of the primal and dual projections.

PROPOSITION 13. *Let* $\mathbf{P}$ *be a closed convex cone and let* $\mathbf{D}$ *be its dual cone. The projection* $\hat{\mathbf{z}}$ *is the unique element of* $\mathbf{P}$ *that satisfies the gradient inequality*

$$(\mathbf{z} - \hat{\mathbf{z}}) \cdot \mathbf{m} \le 0 \quad \textit{for all } \mathbf{m} \in \mathbf{P}.$$

Further, there exists an orthogonal decomposition of $\mathbf{z}$ *into its primal projection and dual projection. That is,*

$$\mathbf{z} = \hat{\mathbf{z}} + \mathbf{e} \quad \textit{with } \hat{\mathbf{z}} \cdot \mathbf{e} = 0.$$

PROOF. [*Exercise.*] First show that $(\mathbf{z} - \hat{\mathbf{z}}) \cdot \hat{\mathbf{z}} = 0$ using the fact that cones contain all rays and that therefore $\|\mathbf{z} - \alpha\hat{\mathbf{z}}\|$ is minimized at $\alpha = 1$. For the gradient inequality, use convexity of the cone and the fact that $\|\mathbf{z} - (1-\varepsilon)\hat{\mathbf{z}} - \varepsilon\mathbf{m}\|$ is therefore minimized at $\varepsilon = 0$. To show that $(\mathbf{z} - \hat{\mathbf{z}})$ is the dual projection $\mathbf{e}$, show that it satisfies the gradient inequality on the dual cone. □

To derive the limiting distribution theory, we will use the elegant geometric description of the conal projection problem given by Fraser and Massam (1989), who apply it to construct an algorithm for the projection. The discussion is therefore limited to convex cones with a certain simple structure. However, the loss in generality is balanced by a set of important insights.

4.4.5. *The dual basis.* Suppose that our convex cone $\mathbf{P}$ is generated as the positive cone induced by a linearly independent set of unit vectors $\mathbf{p}_1, \mathbf{p}_2, \ldots, \mathbf{p}_d$ in d-dimensional space, the primal basis. We can construct a *dual basis* for the dual cone by sequentially finding *unit* vectors $\mathbf{d}_1, \ldots, \mathbf{d}_d$ such that

$$\mathbf{d}_i \cdot \mathbf{p}_j = 0 \quad \text{for } j \neq i.$$

This orthogonality determines $\mathbf{d}_i$ up to its sign, which we determine by making it negatively correlated with its primal partner:

$$\mathbf{d}_i \cdot \mathbf{p}_i \le 0.$$

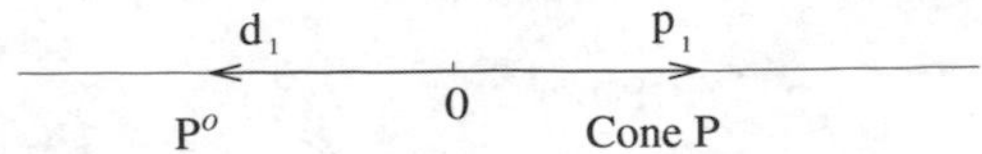

FIG. 4.5. *The dual and primal cones for Example 11.*

It is easy to check that every $\mathbf{d}_i$ is an element of the dual cone and that every positively weighted linear combination of the $\mathbf{d}_i$ is an element of the dual cone. In fact, the dual cone is just the positive cone generated by the dual basis.

Moreover, we can partition the space into 2^d positive cones by using as a basis for each cone $\mathbf{b}_1, \dots, \mathbf{b}_d$, where each $\mathbf{b}_i$ equals either $\mathbf{p}_i$ or $\mathbf{d}_i$. Check that the set of vectors $\mathbf{b}_1, \dots, \mathbf{b}_d$ is also linearly independent. We will call each such a cone a *sector*. We can index the sectors by $\delta = (\delta_1, \dots, \delta_d)$, where δ_i is 1 if $\mathbf{b}_i = \mathbf{p}_i$ and is 0 if $\mathbf{b}_i = \mathbf{d}_i$. The basis vectors used in the sector will be called the *active* basis for that sector. We will denote such a sector by

$$\mathscr{S}(\delta) = \left\{ \sum a_i[\delta_i \mathbf{p}_i + (1 - \delta_i)\mathbf{d}_i] \colon a_i \geq 0 \right\}.$$

In each such sector, the active primal vectors $\mathbf{p}$ are orthogonal to the active dual basis vectors $\mathbf{d}$, a fact that is important in the projection and, later, the distribution theory.

We illustrate these notions with our two simple examples. In Example 11, the cone $\tilde{\mathbf{U}}^*$ is simply a single vector, so we have only a single coordinate z_1 in z space. See Figure 4.5. The induced primal cone is then simply $\{z_1 \geq 0\}$. The projection of v onto this cone is therefore either v itself, if v is nonnegative, or 0, if v is negative. The dual cone is $\{z_1 \leq 0\}$. The reader should check the validity of the theorem in this simple case.

Example 12 has more geometric content. There are now two primal vectors $\mathbf{p}_1$ and $\mathbf{p}_2$. In Figure 4.6, we show the location of the dual basis vectors $\mathbf{d}_1$ and $\mathbf{d}_2$, which are orthogonal to $\mathbf{p}_2$ and $\mathbf{p}_1$, respectively, with direction chosen so as to preserve the negative correlation with the other $\mathbf{p}$ vector. These two bases generate a division of the plane into four conal sectors, corresponding to the primal and dual cones, $\mathbf{P} = \mathbf{P}(1,1)$ and $\mathbf{P}^o = \mathbf{P}(0,0)$, and two regions $\mathbf{P}(0,1)$ and $\mathbf{P}(1,0)$, where we use 0 and 1 to indicate whether the cone was generated with dual or primal basis element in that position.

4.4.6. *Sector decomposition and projection.* The beauty of the sector decomposition is that in each sector there is a simple expression for the projection of $\mathbf{z}$ onto $\mathbf{P}$. Check that in the preceding example,

$$\begin{aligned}
\mathbf{z} \in \mathbf{P} \quad &\Rightarrow \quad \hat{\mathbf{z}} = \mathscr{P}_{\mathbf{P}}\mathbf{z} = \mathbf{z},\\
\mathbf{z} \in \mathbf{P}^o \quad &\Rightarrow \quad \hat{\mathbf{z}} = \mathbf{0},\\
\mathbf{z} \in \mathbf{P}(1,0) \quad &\Rightarrow \quad \hat{\mathbf{z}} = (\mathbf{p}_1 \cdot \mathbf{z})\mathbf{p}_1,\\
\mathbf{z} \in \mathbf{P}(0,1) \quad &\Rightarrow \quad \hat{\mathbf{z}} = (\mathbf{p}_2 \cdot \mathbf{z})\mathbf{p}_2.
\end{aligned}$$

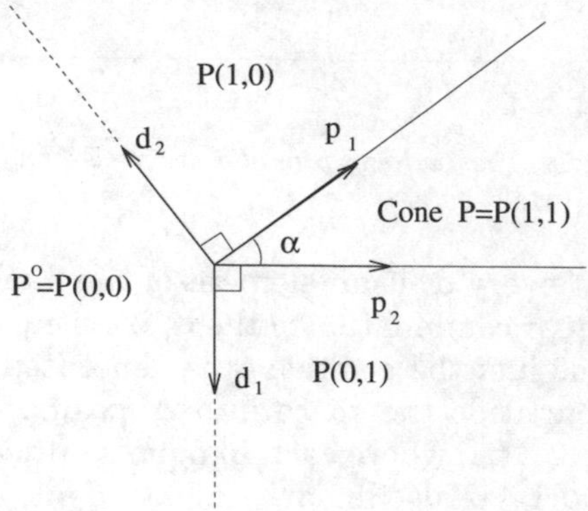

FIG. 4.6. *The cones and sectors for Example 12.*

That is, in this case, and more generally, when $\mathbf{z}$ is in a conal sector, the projection $\hat{\mathbf{z}}$ is formed by *projecting onto the linear space of the active primal vectors*. It is also clear that $\mathbf{e}$ is found in each case by projecting onto the active dual basis for the sector.

It will be useful to characterize the relationships between these projections. Define the linear space $\mathscr{L}_1(\delta)$ to be the linear space spanned by the active primal vectors and let

$$\mathscr{C}_1(\delta) = \left\{ \sum a_i \delta_i \mathbf{p}_i \colon a_i \geq 0 \right\}$$

be the corresponding positive cone. Let $\mathscr{L}_2(\delta)$ and $\mathscr{C}_2(\delta)$ be the corresponding linear and convex cones for the active dual vectors. Let $\mathbf{z}_1$ and $\mathbf{z}_2$ be the linear projections of $\mathbf{z}$ onto $\mathscr{L}_1(\delta)$ and $\mathscr{L}_2(\delta)$, respectively. We then have the following relationships:

PROPOSITION 14. *The following statements are equivalent:* $\mathbf{z} \in \mathscr{S}(\delta)$ *if and only if* $\mathbf{z}_1 \in \mathscr{C}_1(\delta)$ *and* $\mathbf{z}_2 \in \mathscr{C}_2(\delta)$ *if and only if* $\hat{\mathbf{z}} = \mathbf{z}_1$ *and* $\mathbf{e} = \mathbf{z}_2$.

PROOF. [*Exercise.*] It will be useful to note that $\mathscr{L}_1$ and $\mathscr{L}_2$ are orthogonal spaces and that $\mathbf{z}$ has a unique representation in terms of the sector's basis vectors $\mathbf{b}$. □

4.4.7. *The type II LRT.* To do a limiting distribution calculation for the asymptotic version of the likelihood ratio test statistic, namely, $d^2 = \|\mathscr{P}_{\mathbf{P}}\mathbf{Z}\|^2$, we can calculate probabilities via the law of total probability,

$$\Pr(d^2 > t) = \sum_{\text{sectors}} \Pr(d^2 > t | \mathbf{Z} \in \text{sector}) \Pr(\mathbf{Z} \in \text{sector}). \tag{4.11}$$

We tackle this calculation in two steps. Part of the calculation of the limiting distribution turns out to be quite straightforward.

PROPOSITION 15. $\Pr(d^2 > t \mid \mathbf{Z} \in \text{sector}) = \Pr(\chi_k^2 > t)$, *where k is the number of active primal constraints in the sector.*

PROOF. From the preceding proposition, we need to calculate

$$\Pr[\|\hat{Z}\|^2 > \mathbf{t} \mid \mathbf{Z}_1 \in \mathscr{C}_1(\delta) \text{ and } \mathbf{Z}_2 \in \mathscr{C}_2(\delta)],$$

which in turn equals

$$\Pr[\|\mathbf{Z}_1\|^2 > \mathbf{t} \mid \mathbf{Z}_1 \in \mathscr{C}_1(\delta) \text{ and } \mathbf{Z}_2 \in \mathscr{C}_2(\delta)].$$

However, by the independence of $\mathbf{Z}_1$ and $\mathbf{Z}_2$, since they arise from orthogonal projections, the condition on $\mathbf{Z}_2$ is irrelevant in this calculation. Further, the conal structure implies that we can replace the condition $\mathbf{Z}_1 \in \mathscr{C}_1$ with its equivalent condition $\mathbf{Z}_1/\|\mathbf{Z}_1\| \in \mathscr{C}_1$. However, this unit vector has a distribution independent of its length $\|\mathbf{Z}_1\|$ (e.g., by Basu's theorem). Hence our calculation boils down to $\Pr(\|\mathbf{Z}_1\|^2 > \mathbf{t})$, which has the given chi-squared property. □

The second part of the calculation, $\Pr(\mathbf{Z} \in \text{sector})$, is more difficult, and the answer will vary from problem to problem. The problem can be turned into the problem of calculating the surface area of a region on the unit sphere as follows. Let $R = \sqrt{Z_1^2 + \cdots + Z_d^2}$. Because of normality, the conditional distribution of the vector $\mathbf{Z}$ given R is uniform on the sphere of radius $R = r$. It follows that $\mathbf{U} := \mathbf{Z}/R$ is uniformly distributed on the unit sphere. Now $\mathbf{Z}$ is in a positive cone, say $\mathbf{P}$, if and only if $\mathbf{U}$ is in the intersection $\mathbf{I}$ of the sphere with $\mathbf{P}$. The probability of $\mathbf{Z}$ being in the cone is therefore equal to the spherical surface area of $\mathbf{I}$ divided by the total surface area of the sphere. (I will herein refer to "area" rather than "volume," even though the sphere is of arbitrary dimension.) We will illustrate such a calculation shortly.

It follows from (4.11) that the distribution of d^2 has the form of a mixture of chi-squared distributions,

$$\sum_{i=0}^{\dim(\theta)} w_i \chi_i^2,$$

with the weights w_i determined by the probabilities of the vector $\mathbf{z}$ following into various sectors. Such a mixture of chi-squared distributions is called a *chi-bar-squared distribution* and is written $\overline{\chi}^2$.

The chi-bar-squared distributional result holds more generally than for the cones considered here. It can be extended to the positive cones generated by vector sets having an arbitrary number of elements by extending the above ideas. Shapiro (1985) used this fact to show that the chi-bar-squared distribution holds for projections onto an *arbitrary* convex cone. (Essentially, we can approximate arbitrary convex cones with the positive cones associated with a set of primal vectors and then take limits on the number of primal vectors used.)

When a likelihood ratio test has the structure of a convex cone, with the resulting distribution being chi-bar-squared, we will refer to it as a *type II problem*.

4.4.8. *Applications.* Example 11 illustrates the simplest such conal projection result. We have but a single parameter of interest, with inequality constraint on it. The theory above leads us to conclude that the likelihood ratio statistic has the limiting distribution $0.5\chi_0^2 + 0.5\chi_1^2$, where 0.5 is the probability of z_1 being positive, the projection therefore equaling z_1, and the resulting sector distribution being χ_1^2. The second component $0.5\chi_0^2$ arises from z_1 being negative and the projection therefore being $\hat{z}_1 = 0$.

An example of this type from the mixture problem is the following. Suppose wish to test H: $\pi = 0$ versus A: $\pi > 0$ in the mixture model $(1-\pi)f + \pi g$, where f and g are both known. That is, we ask if the distribution f has been contaminated by observations from the second distribution g.

In Example 12, it is clear now that the distribution of d^2 is a mixture of χ_k^2, for $k = 0, 1, 2$. To determine the weights, note first that the weight for the χ_1^2 component is 1/2, because $\mathbf{P}(0,1)$ and $\mathbf{P}(0,1)$ each contribute probability 1/4. (This is the probability of falling in an arc of 90° on the unit circle under the uniform measure on the circle.)

The probability of falling in the other two sectors is proportional to the angles they subtend. It suffices to determine the sector probability for $\mathbf{P}(1,1)$. We need the *angle* α between the primal basis vectors $\mathbf{p}_1$ and $\mathbf{p}_2$, but we need to express it in terms of the original variables that were projected into the $\mathbf{z}$ coordinate system. It can be argued that

$$\cos(\alpha) = \frac{E[\tilde{u}_1(X)\tilde{u}_2(X)]}{\sqrt{E[\tilde{u}_1(X)^2]E[\tilde{u}_2(X)^2]}}, \tag{4.12}$$

in which case the sector probability is $\alpha/2\pi$. (This is the angle between the generating vectors, as expressed using the E_0 inner product. As we have already seen, this agrees with the angle in the z-coordinate system of the transformed vectors.) All the terms on the right hand side of (4.12) can be calculated from the elements of the Fisher information matrix for the parameters at the null hypothesis.

However, there is an important issue here if there are nuisance parameters in the model. In the type I likelihood ratio test theory, there is no possible dependence of the limiting distribution on the nuisance parameters in the null hypothesis. Here, however, there is nothing to prevent the angle α from depending on the value of the nuisance parameter γ under the null hypothesis, in which case the limiting distribution varies over the elements of the null hypothesis.

When this problem arises, we will say that we have a *parameter dependent limiting null distribution*. An example will be given in the mixture problem in the next section.

If parameter dependence occurs, then one must develop a secondary strategy for conducting the test. If one desires the test to have the desired probability of type I error, a conservative strategy would be to use the critical value, say $c(\gamma)$ from the *least favorable* null distribution, with parameter γ. With a little further care as to the asymptotics, one can estimate the nuisance parameter under the null hypothesis, and use the critical value $c(\hat{\gamma})$ from the estimated distribution. In the latter case, one might for the sake of conservativeness employ the least favorable critical value within a confidence interval for the nuisance parameter.

4.5. Asymptotic mixture geometry. Our first problem is to come to an understanding of how we can put the mixture model into the geometric framework of the preceding section. The problem will be that of testing the one-component model against two components. Thus the null hypothesis is described by one-component density $f(t;\phi_0)$. The nuisance parameter in the null hypothesis is the parameter ϕ, which gives us a single nuisance score function $v(\phi,t)$, and this gives us the nuisance score space $\mathbf{V}$.

4.5.1. *Directional score functions.* A *fundamental* difficulty arises in ascertaining the nature of the tangent space $\mathbf{S}$ of score functions at the null hypothesis. In the previous discussion, it was assumed that the score functions under the alternative were also well defined under the null hypothesis. However, this is not true for the usual mixture parameterization. We have three score functions at each alternative parameter value, corresponding to one weight and two component parameters. Their limits as the null hypothesis is approached are problematical, however, because there are problems relating the parameters in the alternative hypothesis to those in the null model.

To illustrate, if we treat the null hypothesis as specifying the equality of the two component parameters, $\phi_1 = \phi_2$, then the score functions for the parameters ϕ_1 and ϕ_2 are both equal to the nuisance score function when evaluated at the null hypothesis. Moreover, the score function for the weights is identically zero there. Thus the score space appears to degenerate to just the score space for the nuisance parameter ϕ.

This analysis is misleading; the easiest way to deal with the problem of determining the score space $\mathbf{S}^*$ is to return to the geometric considerations of Chapter 2. Recall our plots of the structure of the binomial mixture model. The two-component mixture model generates a smooth family of models $\mathbf{F}$ in the probability simplex, a three-dimensional manifold corresponding to the three free parameters. The one-component models are a one-dimensional curve along the edge of this smooth manifold. Recall that our objective is to construct a score manifold that approximates the daggered model surface $\mathbf{F}^\dagger$ in the neighborhood of the null model. We can do this directly as follows. The *directional scores* at a particular null hypothesis point are geometrically constructed by taking limits of the form

$$\varepsilon^{-1}\mathbf{f}_{2\varepsilon}^{\dagger} \longrightarrow \mathbf{s}$$

as $\varepsilon \to 0$, where $\mathbf{f}_{2\varepsilon}$ is a family in $\varepsilon > 0$ of elements of the alternative hypothesis manifold of two-component mixtures that is approaching the null hypothesis point $\mathbf{f}_0$, a one-component density, with sufficient smoothness that the limit exists. We will let the set of all scores created thusly be the *directional score cone* $\mathbf{S}^*$. If indeed we have the simple model structure of the preceding section, where the focal parameter scores are meaningfully defined in the null hypothesis, then this gives the score cone $\mathbf{S}^*$ of that section, which was there defined in terms of the focal and nuisance parameter score functions.

We note first that $\mathbf{S}^*$ is a cone, because one can simply change the definition of the smooth family to alter the speed of the approach to the null hypothesis to get $c\mathbf{s}$. Note, however, that it is not generally true that such a local alternative family can be extended smoothly from positive ε through 0 into negative ε, so that $-\mathbf{s}$ is *not* necessarily in the directional score cone. This is very important in the mixture model, due to the way that the null lies at the boundary of the alternative. It follows that we must pay attention to the sign of the score function if we wish the score surface to approximate the model surface.

4.5.2. *The gradient scores.* This can be further illustrated by considering the scores generated by the weight parameter. Let us consider the boundaries of the parameter space in Figure 4.1 corresponding to $\pi = 0$ or 1. If for each fixed ϕ we construct the score function for the parameter π in the one parameter density

$$(1-\pi)f(t;\phi_0) + \pi f(t;\phi),$$

then taking its limit as $\pi \to 0$, we obtain a score function of the form

$$s_\phi(\phi_0;t) = \frac{f(t;\phi)}{f(t;\phi_0)} - 1.$$

This leads to a number of important observations:

1. This generates an *infinite* family of directional score functions $\mathbf{C} = \{c\mathbf{s}_\phi\}$ corresponding to the infinitely many possible values of ϕ. By considering this family, and certain limits, we will be able to determine all the local corrected score functions. However, the dimension of the score space is no longer equal to the number of parameters in the alternative, and this will prevent us from using the simple geometry of the preceding section to derive the limiting distributions.
2. The sample version of the above score is

$$n\sum d(t)s_\phi(\phi_0;t) = D_{\phi_0}(\phi). \tag{4.13}$$

   That is to say, the unicomponent gradient function represents the collective sample values of these scores. For this reason, we will call $\mathbf{s}$ a *gradient score*.
3. These scores are one directional in that the one parameter family used to generate them does not extend beyond $\pi = 0$, so the score is not two-sided. That is, the model surface in the simplex has an edge at the null hypothesis.

4. The gradient scores are related very simply to the dagger operation. That is, we have

$$\mathbf{s}_\phi = \mathbf{f}_\phi^\dagger.$$

Thus if we wish to picture the cone generated by these scores, we plot the one-component model in the dagger space and the cone $\mathbf{C}$ is all rays from the origin through other points in the model surface.

Before considering further issues, we ask the reader to solidify his or her understanding by considering how these facts relate to the pictures of the binomial mixture models found in Chapter 2. The pictures are roughly the same, with the coordinates undergoing some stretching by the daggering operation. Thus, using the Bin$(2, p)$ as an example, Figure 4.7, we see that the gradient score vectors $\mathbf{s}_\phi$ correspond to all the vectors from the origin (the null model) to points on the curve $\{\mathbf{f}_\phi^\dagger\}$ generated by the one-component model.

From the figure we also obtain an important insight. We see that there are sequences of two-component models that approach the null model in such a way that the directional score is not in $\mathbf{C} = \{c\mathbf{s}_\phi\}$. Thus we must create other score functions if we are to generate the entire directional score tangent cone $\mathbf{S}^*$.

4.5.3. *Other directional scores.* We return to our derivation of the $C(\alpha)$ test in order to generate some further score functions. We follow the Neyman–Scott derivation, letting the distribution G be a two point distribution, in which case the location-scale family $\Phi = a + b\Theta$ is also entirely two point distributions. If we let $b \searrow 0$, we should expect to find that the directional score for any such two-component alternative results in the dispersion score $v_2(\phi_0, t)$. However, we note that because the first derivative in b is zero, we must let $b = \sqrt{\varepsilon}$ to get this score. Moreover, we get exactly the same score (no change in sign)

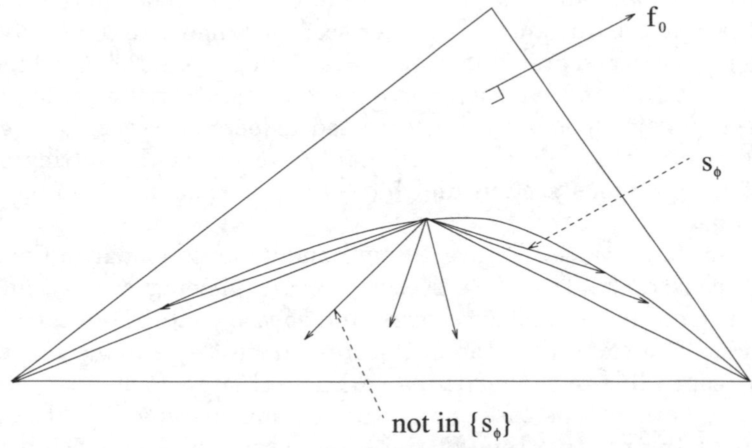

FIG. 4.7. *The directional scores for the* Bin$(2, p)$ *model.*

if $b = -\sqrt{\varepsilon}$. That is, v_2 generates a ray in only one direction in the tangent cone. In the plot for the model $\mathrm{Bin}(2, \phi)$, it corresponds to a ray down into the two-component models.

A further set of scores can be generated if we consider all approaches to the null model in which $a \to \phi_0$ at the same time as $b \to 0$, with appropriate rates, in which case we get scores of the form $c_2\mathbf{v}_2 + c_1\mathbf{v}_1$, where $c_2 \geq 0$.

At this point, a rather delicate analysis is needed to verify that we have found *all* possible directional score functions. We offer some heuristics for this: If the local family of alternatives is such that π does not go to 0 or 1, we must have ϕ_1 and ϕ_2 converging to ϕ_0. In this case, we will get a limit involving the dispersion score and the nuisance score. If π does converge to 0, then ϕ_1 must converge to ϕ_0 and the resulting score will be a gradient score provided ϕ_2 converges to something other than ϕ_0.

As in the preceding section, the relevant portion of the directional score cone for determining the asymptotic distributions is orthogonal to the nuisance parameter score space. This leads to a further simplification in the analysis, because, fortunately, the dispersion score does not need to be handled separately from the gradient scores $\mathbf{s}_\phi$ in the geometric analysis. That is, we claim that

$$\tilde{\mathbf{S}}^* = \tilde{\mathbf{C}}.$$

This is because the relevant scores for asymptotic analysis are the corrected scores obtained by the E_0 regression residuals $\tilde{\mathbf{s}}_\phi = \mathbf{s}_\phi - \rho\mathbf{v}_1$. Noting that $\mathbf{s}_\phi/(\phi - \phi_0) \to \mathbf{v}_1$ as $\phi \to \phi_0$, the interested reader should check that the normalized dispersion score $\tilde{\mathbf{v}}_2/\|\tilde{\mathbf{v}}_2\|$ is the limiting vector of the normalized corrected gradient scores $\tilde{\mathbf{s}}_\phi/\|\tilde{\mathbf{s}}_\phi\|$ as $\phi \to \phi_0$.

Thus we only need consider the corrected score cone $\tilde{\mathbf{C}}$ generated by the closure of the cone of the corrected gradient scores, because the other scores are limit points thereof.

4.5.4. *Simple binomial examples.* In the $\mathrm{Bin}(2, p)$ plot, there are only two dimensions, and the nuisance parameter score is tangent to the unicomponent model, so the corrected gradient score space $\tilde{\mathbf{C}}$ must lie in the one-dimensional subspace orthogonal to it. It is a directional cone, pointing down into the model. This direction corresponds to the corrected dispersion score $\tilde{\mathbf{v}}_2 = \mathbf{v}_2 - \rho\mathbf{v}_1$. Thus, after this reduction of dimensionality, we are in the setting of Example 11 of the previous section and have the corresponding chi-bar-squared distribution.

In the $\mathrm{Bin}(3, p)$ model, Figure 4.8 will enable us to visualize the relevant geometric constructs. The cross section we are viewing corresponds to the plane orthogonal to the nuisance score and so will contain the corrected gradient scores. The corrected gradient scores all lie *between* the two extremal scores corresponding to the mixtures with latent support at $\theta = 0$ and $\theta = 1$, respectively. Thus in this case the corrected tangent cone $\tilde{\mathbf{C}}$ for the model is generated as the positive cone of these two extremal scores. Thus, because of the limited number of dimensions in the simplex, the corrected tangent

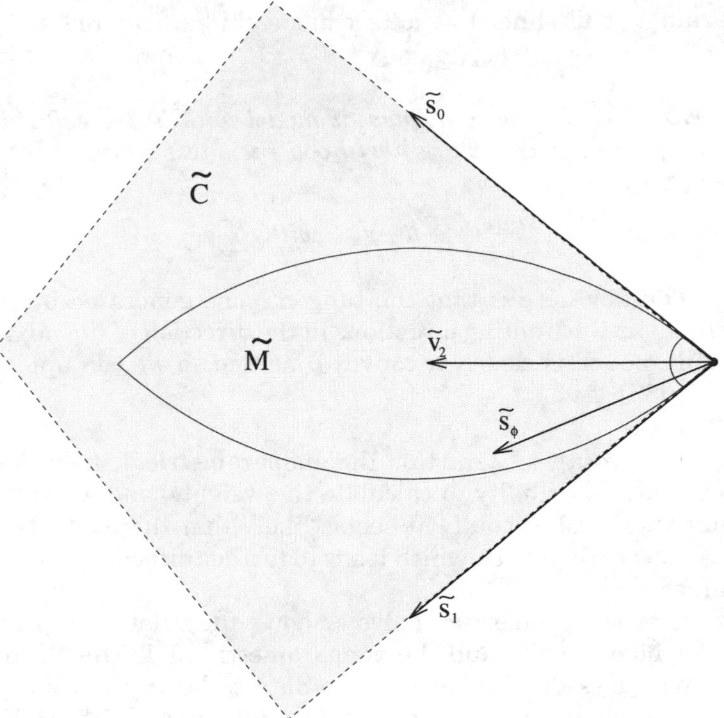

FIG. 4.8. *The corrected gradient score cone of the* Bin(3, p) *model.*

score space is just two dimensional, despite the fact that it was generated by infinitely many gradient score functions. The corresponding picture from our conal analysis is Figure 4.6.

Thus in both these binomial models, the geometry coincides exactly with the type II LRT theory of the previous section. We can conclude that in a Bin$(2, p)$ model, the test of one versus two components has the chi-bar-squared distribution $0.5\chi_0^2 + 0.5\chi_1^2$. In the Bin$(3, p)$ distribution, we have a mixture of chi-squared $(0.5-\alpha)\chi_0^2+0.5\chi_1^2+\alpha\chi_2^2$, where the component weights must be determined by calculation of α in (4.12). This calculation can readily be carried out and we find that the angle is not constant. For example, $\alpha = 0.167$, 0.193 and 0.226 when the log odds parameter equals 0, 2 and 4, respectively.

This proves that in this binomial case the LRT distribution is *parameter dependent* in the null hypothesis, and suggests that it is unlikely to be distribution constant in many other examples.

4.5.5. *The nonparametric LRT.* Before proceeding to the next level of difficulty, where the chi-bar-squared distributions fail, we can use what we already know to make an observation about the nonparametric likelihood ratio test, because the necessary geometric background has been laid. The result is that the nonparametric likelihood ratio test has a chi-bar-squared distribution as well.

PROPOSITION 16. *In the multinomial model with $T+1$ cells, under regularity the nonparametric mixture likelihood ratio test statistic has a limiting distribution of the form*

$$w_0\chi_0^2 + \cdots + w_T\chi_T^2 \quad \textit{with} \sum w_j = 1.$$

PROOF. The key here is that the tangent cone generated by the mixture models, viewed as the limiting directions in the direction of the mixture models from the null model, is clearly a convex cone and so we can apply the result of Shapiro (1985). □

Although this result is a start on the nonparametric distribution theory, it is useless without the ability to calculate the weights, and we anticipate that is a difficult issue. Additionally, it seems likely that there will be parameter dependence in the weights w, which leads to further difficulties in constructing critical values.

We note that the geometric analysis shows the relationship between the $C(\alpha)$ test for homogeneity and the nonparametric LRT. The Neyman dispersion score measures the tendency of the data to lie in the central direction $\mathbf{v}_2$ of the cone, but ignores the more subtle features of the conal structure. However, it has the clear advantage of an easy distribution theory and simple calculation. It seems likely that one can develop extensions of this test that measure some additional departure in the direction of heterogeneity, such as skewness features, yet still retain manageable limiting distributions.

4.5.6. *A nonconvex score cone.* Once we leave the three-dimensional simplex, where the corrected score space is forced to live in two dimensions, the distribution of the likelihood ratio test for one versus two components becomes significantly more challenging.

Suppose the one-component model is $\text{Bin}(4, p)$. From our earlier geometric analysis we know that the set of mixture probability vectors is a four-dimensional set and that if we constrain ourselves to the plane in the simplex orthogonal to the nuisance score function, then the two-component models generate a two-dimensional surface in the three-dimensional space: see Figure 4.9. In fact, there are two boundary surfaces to the mixture set consisting of the two types of mixtures with index 2: those that mix two values of θ that are in $(0,1)$ and those that mix $\theta = 0$, $\theta = 1$ and one $\theta \in (0,1)$. Between these two surfaces are two seams, one on each side, consisting of the mixtures of index 1.5. Only the first surface of index 2 concerns us, because the other would be considered, in a statistical sense, to consist of three-component models.

We can imagine the plot of the surface as being like the $\text{Bin}(3, p)$ plot, only with an extension into the third dimension; something like an American

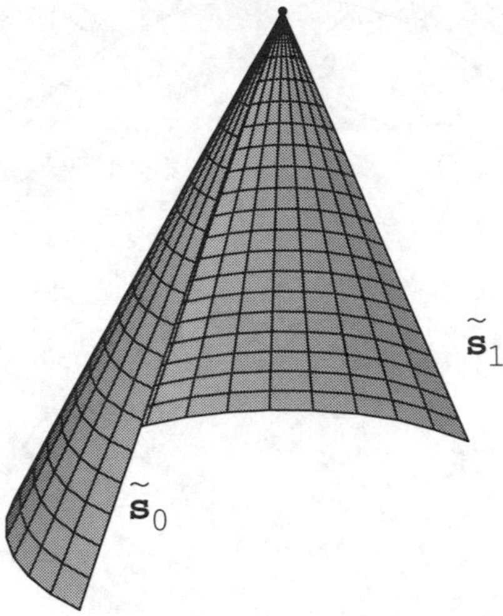

FIG. 4.9. *The corrected gradient score cone of the* Bin(4, *p*) *model.*

football, but with only two seams. The null model sits at the top of the football, so we can picture those tangent vectors to the surface at the null model that correspond to the directions toward two-component models as creating a cone that has the general shape of a half tepee, sliced in two from its apex down.

That is, this is a situation where the tangent cone $\mathbf{S}$ is *not* convex, and so we cannot use the theory of the type II likelihood ratio test. In essence, the two-dimensional score manifold does not lie in a two-dimensional plane.

Of importance in our further discussion will be the two extremal vectors, corresponding to the two edges of the tepee surface. These correspond, in the Bin$(4, p)$ case, to the two seams with index equal to 1.5, and so arise from a mixture with $p = 0$ on one edge and $p = 1$ on the other.

4.6. The LRT on nonconvex cones. As part of our analysis, we must therefore gain some further understanding of projections onto curved surfaces. In addition, the limiting distribution theory will be closely related to results for normal theory nonlinear regression, where exact results are unusual.

4.6.1. *Projections onto nonconvex cones.* We first consider the issue of the nonuniqueness of projections. If we are in $\mathbf{R}^2$—the Euclidean plane—and we wish to find the projection of a point $\mathbf{z}$ onto a curve $\mathscr{C}$, such as a parabola or hyperbola or circle, then there may be multiple points of minimum distance from $\mathbf{z}$ on the given curve. To take an extreme, but instructive, case, if the curve $\mathscr{C}$ is the unit circle and $\mathbf{z}$ is the center of the circle, then it is equidistant from *all* the points of the circle. However, in every other case in this same example, there is a unique projection of $\mathbf{z}$. For any point $\mathbf{z}$ that is outside $\mathscr{C}$, this is clear. On the other hand, if $\mathbf{z}$ is inside the circle, the set of points that have a fixed distance c from $\mathbf{z}$ lie on a circle $\mathscr{S}_c$ of radius c. See Figure 4.10. As c grows, we

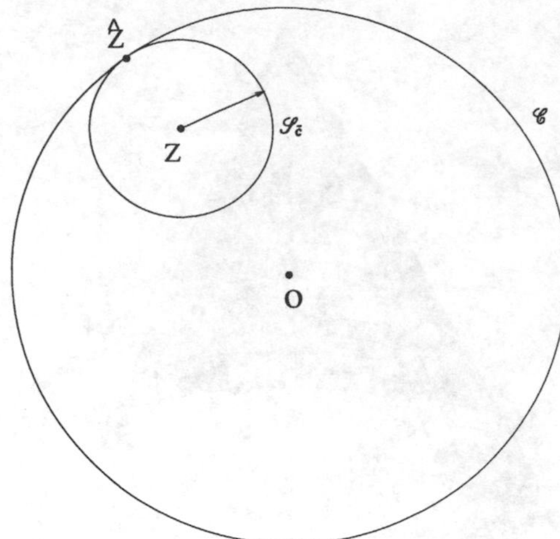

FIG. 4.10. *Projections onto a circle.*

can visualize that the minimum distance point on $\mathscr{C}$ corresponds to the first intersection point of $\mathscr{C}$ and $\mathscr{S}_{\tilde{c}}$, where $\tilde{c}$ is the minimum possible distance. This intersection point is unique due to the greater curvature of $\mathscr{S}_{\tilde{c}}$ than $\mathscr{C}$. Thus the curvature of the surfaces involved play a critical role in uniqueness considerations, as well as the multimodality of the distance function.

Although the projections $\hat{\mathbf{z}}$ on $\mathscr{C}$, may not be unique, there are still some useful facts available if $\mathscr{C}$ is a closed cone.

1. The distance function is continuous, so for closed surfaces (containing their limit points), there does exist a well defined minimum distance to the cone $\|\mathbf{z} - \hat{\mathbf{z}}\|$ that is attained for some point $\hat{\mathbf{z}}$.
2. Since we are projecting onto a *cone*, whatever solution we find, say $\hat{\mathbf{z}}$, is orthogonal to $\mathbf{z} - \hat{\mathbf{z}} := \mathbf{e}$, so we still have the basic orthogonal decomposition into fitted values and residuals,

$$\mathbf{z} = \hat{\mathbf{z}} + \mathbf{e},$$

with the accompanying sums of squares decomposition, even if we no longer have uniqueness for the vectors in this decomposition or a dual projection interpretation of the residual $\mathbf{e}$.

To visualize the concepts in the second point, imagine that the cone consists of two rays from the origin in $\mathbf{R}^2$. Then when $\mathbf{z}$ is between the two rays, it can be projected onto either ray to find a point of local minimum distance, and when $\mathbf{z}$ is exactly midway between, the two local minimum distances are equal. See Figure 4.11. However, in both cases $\hat{\mathbf{z}}$ is orthogonal to $\mathbf{e}$. Because of this, we may continue to use the relationship

$$\|\mathbf{z}\|^2 = \|\hat{\mathbf{z}}\|^2 + \|\mathbf{e}\|^2. \tag{4.14}$$

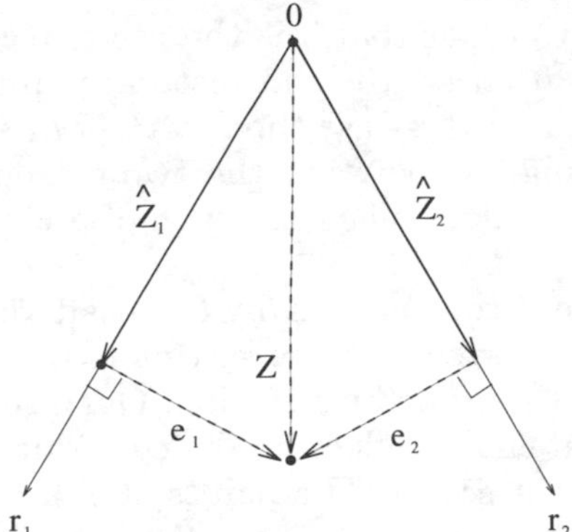

FIG. 4.11. *Nonunique projections onto two rays, with unique lengths.*

Another useful fact is that we can write an explicit formula for $\|\hat{\mathbf{z}}\|^2$ based on the elements of $\mathscr{C}$. We claim that

$$\|\hat{\mathbf{z}}\| = \sup\left\{\frac{(\mathbf{z}\cdot\mathbf{m})^+}{\|\mathbf{m}\|}\colon \mathbf{m} \in \mathscr{C}\right\}. \tag{4.15}$$

The argument goes as follows: View each ray $\{c\mathbf{m}\colon c \geq 0\}$ as a cone on which we do a projection, arriving at $\hat{\mathbf{z}}$ and $\mathbf{e}$ that depend on $\mathbf{m}$. If we do so, we find that the length of "$\hat{\mathbf{z}}$" is $(\mathbf{z}\cdot\mathbf{m})^+/\|\mathbf{m}\|$. Now, for each ray $\mathbf{m}$ the decomposition (4.14) holds, and our goal in projection is to minimize the residual term $\|\mathbf{e}\|^2$. However, because $\|\mathbf{z}\|^2$ is fixed, we can equivalently maximize $\|\hat{\mathbf{z}}\|^2$. The result (4.15) now follows.

4.6.2. *Measuring distances.* Another important issue arises when the model's corrected directional score cone is not convex. We will need to determine statistically appropriate ways to measure distances along the cone. An important parameter in our statistical analysis will be an arc length distance along the unit sphere in the appropriate metric. We will derive it here, in advance of the main result.

We set up the appropriate geometry. In the preceding section, we transformed from the E_0 geometry appropriate to the scores into a z-coordinate system, because in this space everything could be recognized as the projection of standard normal variables onto convex cones. The arguments could have been worked out directly in the dagger space, but the sense of simplification would have been lost. We presume we are working in such a transformed space, and if we refer to a particular score functions, we are referring to their coordinate representation.

If we imagine our (transformed) corrected score cone, with the half-tepee shape, with apex at the origin, then it intersects the unit sphere in a one-dimensional curve Γ on the surface of the sphere. (The unit sphere in the E_0 geometry consists of mean zero variables of variance 1.)

If the cone is actually *flat*, so that it is the positive cone of its two extremal vectors, then Γ is a *great circle*, the shortest path on the sphere between its two endpoints. Moreover, in this case the *length of this path* between the two endpoints equals the angle α between the two extremal rays at the origin, because we are merely walking along the rim of a circle that connects these rays.

However, if the tangent cone is actually the positive cone generated by the two extremal scores, then we are in the setting of the type II likelihood ratio test and the results of the preceding section. Our problem arises due to the fact that Γ is not a great circle, reflecting the curvature of the tangent surface generated by the gradient scores. The points of Γ are the coordinates of the normalized (to variance 1) corrected scores

$$g_\phi(t) := \tilde{s}_\phi(t)/\sqrt{E[\tilde{s}^2_\phi(X)]}.$$

As ϕ varies, $\mathbf{g}_\phi$ traces out the curve Γ on the unit sphere. See Figure 4.12.

We will need the length of this curve in the z-coordinate sense. To solve this, we transform back into the original coordinates and find the E_0 length around the sphere. To do this, one can break the parameter space into a grid of intervals, say $\phi_i < \phi_{i+1}$, and sum the secant distances

$$d_i = \sqrt{E[g_{\phi_{i+1}}(X) - g_{\phi_i}(X)]^2}$$

between neighbors $\mathbf{g}_{\phi_{i+1}}$ and $\mathbf{g}_{\phi_i}$ to arrive at the appropriate approximating sum

$$\text{arc length} \approx \sum_i \sqrt{E[g_{\phi_{i+1}}(X) - g_{\phi_i}(X)]^2}.$$

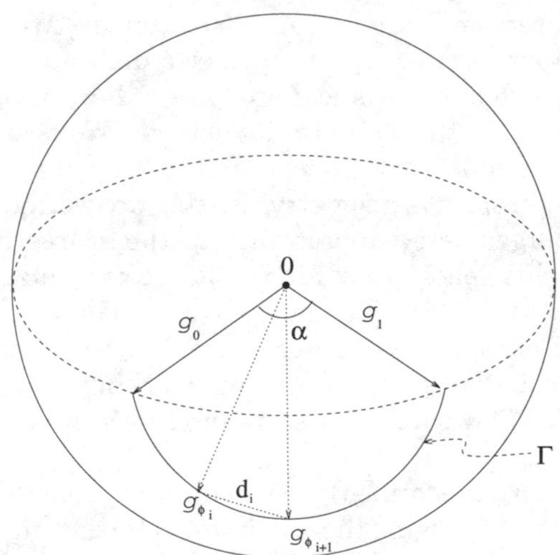

FIG. 4.12. *The trace of the curve* Γ *on the unit sphere.*

If we let the partition of the parameter space grow finer, we see that this approximation converges to

$$\text{Arc}(\Gamma) = \text{arc length} = \int \sqrt{E\left[\frac{\partial}{\partial \phi} g_\phi(X)\right]^2}\, d\phi. \tag{4.16}$$

[If we calculate the arc length using a partition approximation, somewhat greater accuracy can be obtained by summing the arc lengths α_i on the sphere rather than secant distances d_i, using the relationship $d_i^2 = 2 - 2\cos(\alpha_i)$.]

We have carried out these calculations, using the grid approximation, for several binomial models. In Table 4.1, we show these calculations. The parameter ϕ is the natural parameter; that is, the log odds. For comparison, we have shown the great circle distance between the two extremal rays of the cone so as to show that the statistical curvature in the problem can significantly increase the surface of the tangent score cone. In the binomial model, the arc length goes to infinity as $N \to \infty$. In the next section we will find how the arc length shows up in the limiting distribution.

4.6.3. *Tubes and distributions.* Now we tackle the rather severe distributional problem. Our first reduction is to turn this into a problem involving the calculation of surface areas on the unit sphere. The random variable $U = \|\mathbf{e}\|/R = \|\mathbf{e}\|/\|\mathbf{z}\|$ is scale invariant and so is statistically independent of R by Basu's theorem. Thus we can find its distribution by calculation of its conditional distribution given any fixed value of R, which we will take to be $R = 1$. However, the statistic whose distribution we desire is $\|\hat{\mathbf{z}}\|^2 = (1 - U^2)R^2$, so if we calculate the distribution of U, then we can use independence and the known chi-squared distribution of R^2 to find the desired distribution.

Now we use the fact that $\mathbf{Z}$ is, conditionally on $R = 1$, uniformly distributed on the unit sphere to note that the probability $\Pr[U \leq u]$ is the volume of a *tube* about the curve Γ. In Figure 4.13 we have attempted to recreate the geometry of the situation. If we let θ be in $[0, \pi/2]$ and consider all points that are within arc length θ of a point $\mathbf{g}_\phi$ on Γ, then we have a spherical cap at that point. Points $\mathbf{z}$ that are in that cap have a projection distance $\|\mathbf{e}\|$ to the cone no larger than $\sin\theta$.

TABLE 4.1

N	ϕ	Arc/2π	$\alpha/2\pi$
3	0	0.167	0.167
3	2	0.193	0.193
3	4	0.226	0.226
4	0	0.240	0.206
6	0	0.363	0.236
20	0	0.887	0.250

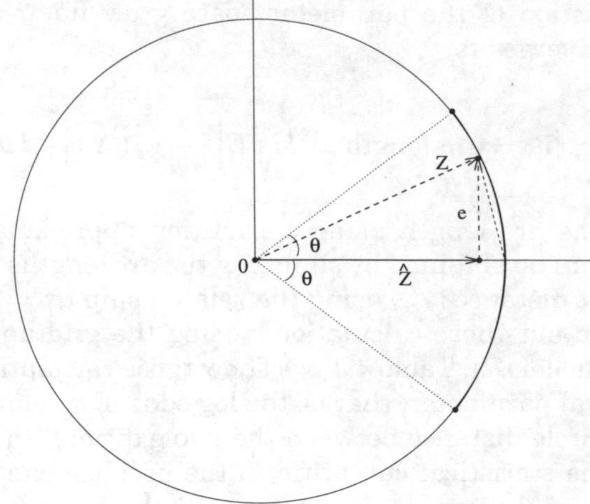

FIG. 4.13. *A cross section of the spherical cap with angular radius θ.*

Thus if we find the spherical volume of the set of all points within θ in arc length from Γ, and divide it by the total surface "volume" of the sphere, we will have the probability that $U \leq \sin(\theta)$.

The set $\mathscr{T}_\theta := \{\mathbf{z}\colon U \leq \sin\theta, R^2 = 1\}$ is the *tube of radius θ about the curve* Γ. It is sketched in Figure 4.14, showing in particular that the two endpoints of Γ generate two semispherical caps.

We first carry out the calculation under the simplifying assumption that the curve Γ lies on a great circle (geodesic). We assume that the angle α it subtends is less than π, so that curve wraps no more than half way about the sphere. We let $V_1, \ldots, V_d$ be the uniformly distributed coordinates of the sphere and we suppose that we have rotated the sphere about so that the great circle is

$$\{(v_1, v_2, 0, \ldots, 0)\colon v_1^2 + v_2^2 = 1\}.$$

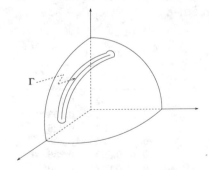

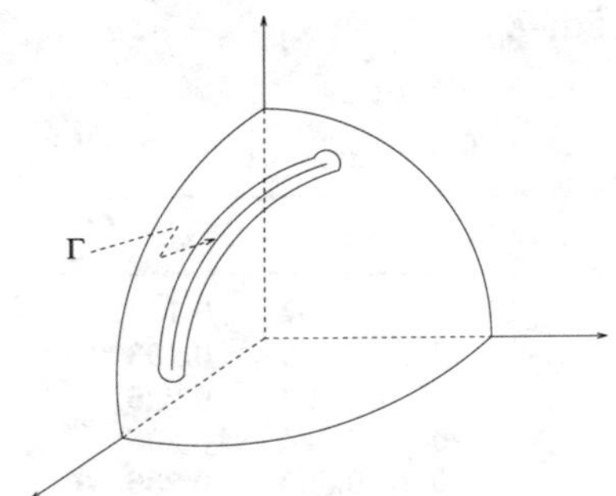

FIG. 4.14. *A tube of radius θ about the curve* Γ.

In this situation, we have already calculated the probability distribution for the likelihood ratio statistic, because we are in the setting of the type II likelihood ratio test, with the geometric structure of Example 12. The corrected score tangent cone is generated as the positive cone of the two extremal rays of the curve Γ, so it lies in the space $\{(z_1, z_2, 0, \dots, 0)\colon z_1 \in R,\ z_2 \in R\}$ and so we can ignore the orthogonal variables $Z_3, \dots, Z_d$, here corresponding to the spherical variables $V_3, \dots, V_d$.

We rederive the distribution for Example 12 by starting with the probability that the uniform sphere variable $\mathbf{V}$ falls in the tube. Provided that the angle θ is in the range $[0, \pi/2)$, so that the tube does not overlap itself, an elementary geometric argument shows that

$$\text{(4.17)}\qquad \Pr[U < \sin\theta] = \Pr[V_1 > \cos\theta] + \frac{\alpha}{2\pi}\Pr\left[\sqrt{V_1^2 + V_2^2} > \cos\theta\right].$$

The first term in this expression is the probability of falling in the two endcaps, which together make one complete semispherical cap. [Note that here V_1 refers not to the first coordinate, but the distribution of the first coordinate]. The second term comes from the body of the tube, excluding the endcaps. By substituting $w = 1 - \sin^2\theta = \cos^2\theta$, this last formula becomes

$$\text{(4.18)}\qquad \Pr[1 - U^2 > w] = \frac{1}{2}\Pr[V_1^2 > w] + \frac{\alpha}{2\pi}\Pr[V_1^2 + V_2^2 > w],$$

where the factor 1/2 arises from the symmetry of V_1's distribution about zero. From here, we may finish the calculation of the distribution by noting that $R^2 \cdot (V_1^2 + V_2^2)$ has a χ_2^2 distribution and $R^2 \cdot V_1^2$ has a χ_1^2 distribution, so an easy *exercise* shows

$$\text{(4.19)}\qquad \Pr[\|\hat{\mathbf{Z}}\|^2 > t] = \frac{1}{2}\Pr[\chi_1^2 > t] + \frac{\alpha}{2\pi}\Pr[\chi_2^2 > t],$$

exactly in accordance with our earlier derivation of the type II likelihood ratio test in Example 12.

4.6.4. *Approximations for tubes.* Hotelling (1939) showed that the tube formula in (4.17) is still an equality if the curve is not a great circle, provided that the curve is regular, and the chosen angle θ is sufficiently small and one replaces α with the arc length $A(\Gamma)$ of the curve. However, equality does fail for large values of θ when the tube displays curvature, that is, when Γ is not a great circle curve, and this failure occurs when θ exceeds the smallest value of the spherical radius of curvature of Γ.

However, Naiman (1986) has shown that even when equality fails to hold, the two sides of (4.17) are always related by an inequality of the form $\leq$. Heuristically, the inequality arises because when θ grows sufficiently large, the curvature of Γ causes a kink in the tube, in which case it has *less* volume than predicted by the formula on the right-hand side.

This said, we will say that we have a *type III likelihood ratio problem* if the tangent cone of the corrected scores is two dimensional and the corresponding curve Γ is smooth, with no points of self-intersection.

PROPOSITION 17. *In any type III likelihood ratio problem with finite arc length $A = A(\Gamma)$, the limiting distribution of $\|\hat{\mathbf{Z}}\|^2$ satisfies*

$$\Pr[\|\hat{\mathbf{Z}}\|^2 > t] \leq 0.5 \Pr[\chi_1^2 > t] + \frac{A}{2\pi} \Pr[\chi_2^2 > t].$$

Moreover, the ratio of the two sides goes to 1 as $t \to \infty$, provided that the curve Γ has spherical curvature that is bounded above.

PROOF. To prove the inequality, we apply Naiman's result to (4.17) and follow its consequences through (4.18) and (4.19). For the second part, we note that as long as θ is sufficiently small that the tube inequality is an equality, the first bound is an equality; hence, so is the second upper bound for w sufficiently close to 1. This implies that the upper tail of $\|\hat{\mathbf{Z}}\|^2 = R^2(1 - U^2)$, which is determined by large values of R and $W := (1 - U^2)$, is asymptotic to the bound, by the following line of argument. Let G be the distribution function for W and let a be such that the bound is exact for $W \geq a$. We can show that the tail bound is exact if the following ratio converges to zero as $t \to \infty$:

$$\frac{\int_0^a \Pr\{R^2 > t/w\}\, dG(w)}{\int_a^1 \Pr\{R^2 > t/w\}\, dG(w)} \leq \frac{\int_0^a \Pr\{R^2 > t/w\}\, dG(w)}{[1 - G(a)] \Pr\{R^2 > t/a\}}.$$

Thus it suffices to show that

$$\frac{\Pr\{R^2 > t/w\}}{\Pr\{R^2 > t/a\}} \to 0,$$

for $w < a$, since the ratio being less than 1 implies dominated convergence. However, this last statement is true by an application of l'Hôpital's rule and the use of the appropriate χ^2 density. □

Although it is perhaps possible to get more accurate descriptions of the limiting distribution theory, it is unlikely that it can be done without considerably more effort. However, such a geometric study will no doubt have further payoffs in understanding the nature of the problem of multimodal likelihoods. Note also that we can accommodate multinomial models with auxiliary parameters in this analysis; they will show up as nuisance scores in the calculation of the corrected mixture scores. Finally, we note that the limiting distribution may not be operative if the sample size is small relative to the parameter space, as we discuss in the next section.

4.6.5. *The arc length problem.* Despite the fact that there is much work to be done to turn the preceding theory into a viable strategy in the wide range of strictly multinomial problems that it could be applied to, it is hard to avoid the temptation to turn it to use in understanding the much studied mixture problem of testing two normal components. We will consider here just the case of two normal components with common known variance, because the unknown variance case considerably changes the geometry.

To enhance our understanding, we imagine that the problem has been discretized, so that we are in a situation similar to a $\text{Bin}(N, p)$ with N large. There are several important features of this problem that we can now see more clearly.

First, as the binomial parameter N increases, the arc length parameter grows without bound, increasing the weight applied to the χ_2^2 component in the approximation. Thus the inequality says that the likelihood ratio test is gaining heavier and heavier tails. If one carries out the formal calculation of the arc length using (4.16) in the normal model, one finds in fact that it becomes infinite when integrated over the parameter space.

This calculation confirms Hartigan's conclusion that the likelihood ratio test statistic diverges to infinity. However, in addition to not being useful, this result appears in contradiction to the great stability that has been found in many simulation studies. Indeed, the simulation study of Böhning, Dietz, Schaub, Schlattman and Lindsay (1994) shown in Figure 4.15 shows that for a sample of size 10,000, the distribution appears very much like the chi-bar-squared approximation, with some small finite arc length. (*Note:* The plot is of the conditional CDF of the nonzero statistic values only, because our right tail approximation gives no information about the probability of a zero.)

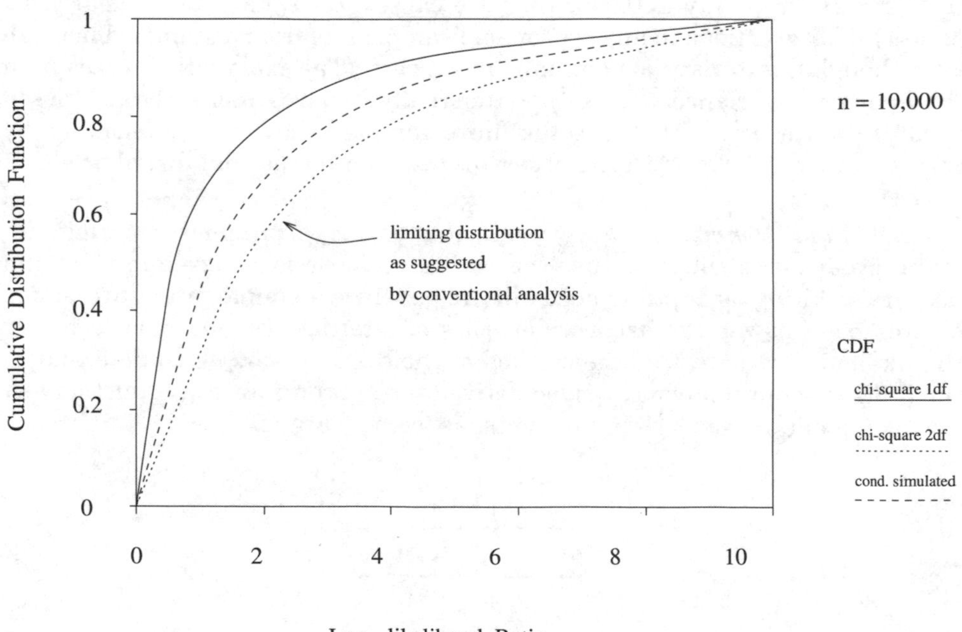

FIG. 4.15. *A simulation study of the likelihood ratio test distribution.*

To explain this phenomenon, suppose we were to restrict the search for a two-component alternative to a finite interval of parameter values, say $\phi \in [L, R]$. Then the theory of the preceding section would lead us to identical conclusions, but the arc length would be calculated over the restricted interval; call this length Arc$[L, R]$.

The importance of this is that in practice, for any finite sample size, we do not need to search for the two-component solution over an infinite range. For example, in a two-component normal mixture problem, there can be no likelihood solutions for which the support points are not between the smallest and largest order statistic of the data, $y_{1:n}$ and $y_{n:n}$. Thus a more relevant calculation would seem to be the arc length $A_n = A[L_n, U_n]$ corresponding to $L_n = E[Y_{1:n}]$ and $U_n = E[Y_{n:n}]$. In the normal model, this *effective arc length* grows very slowly in n. Some values are given in Table 4.2.

Although this gives us some further guidelines as to what to expect in a distributional theory, there are also some technical problems with the asymptotics that one should consider. The distributional theory relies on the corrected scores being asymptotically normally distributed. However, in the mixture case, the gradient scores $\tilde{s}_\phi$ can be a long way from achieving their limiting distribution for any finite sample size. Consider the normal case, where the selected null model is $N(0, 1)$:

$$s_\phi(t) = \frac{\exp(-0.5(t-\phi)^2)}{\exp(-0.5t^2)} - 1 = \exp(\phi t - 0.5\phi^2) - 1.$$

If ϕ is very large, say 100,000, then we can expect $s_\phi(X_i)$ to be nearly -1 across all observations in any reasonable sample size from a standard normal, even though the statistic is nominally mean zero. (The explanation arises from the enormous variance.) Thus the sum really behaves more like a Poisson variate than a normal because the finite sample behavior is determined by rare events, namely, very large observations from the normal distribution.

4.6.6. *Final comments.* As we indicated earlier, a pragmatic strategy for many problems is simply to use the Neyman dispersion score approach and use the resulting asymptotic normality to generate a simple procedure. If one desires more power against a wider class of alternatives, one must consider the likelihood ratio test as the method of choice. However, we do note that in the process of our analysis, we also derived the appropriate score function test for testing one versus two components, as we now argue.

TABLE 4.2

n	$A_n/2\pi$
44	0.44
740	0.75
31,000	1.07
3,500,000	1.40

Recall from (4.13) that inner product of the directional scores $\mathbf{s}_\phi$ from null value ϕ_0 with the data vector resulted in the normalized gradient $n^{-1}D_{\phi_0}(\phi)$. If we transform the projection statistic $\|\hat{\mathbf{z}}\|^2$ from (4.15) back into the original model geometry, we see that the projection statistic, and therefore the likelihood ratio statistic, is asymptotically equivalent to the *generalized score statistic:*

$$U^2(\phi_0) := n^{-1/2} \sup_\phi \left\{ \frac{D^+_{\phi_0}(\phi)}{\|\tilde{\mathbf{s}}_\phi\|} \right\}^2 .$$

This statistic may be estimated under the null hypothesis by $U^2(\hat{\phi})$. This statistic will have the same asymptotic distribution as the likelihood ratio test and is considerably easier to compute provided that the corrected score variances can be calculated explicitly. We note that the argument of the supremum

$$\frac{D^+_{\phi_0}(\phi)}{\|\tilde{\mathbf{s}}_\phi\|}$$

approaches the positive part of the Neyman dispersion test statistic as $\phi \to \phi_0$, so the generalized score test clearly uses wider properties of the gradient than the Neyman test.

There is also a score test corresponding to the likelihood ratio test for one component against an arbitrary number of components, but it is considerably more complicated to compute.

Finally, we note that for both the likelihood ratio test and the generalized score test, the arc length problem goes away provided that one is willing to group data in the tails of the distribution. In a standard contingency table analysis one can group cells together so as to improve asymptotic approximations, and one can do so here. One can either group the data throughout its range, making sure the bins in the tails have sufficient observations, or one can construct a likelihood from the densities in the middle, but use the appropriate distribution functions in the tails. In either case, the arc length calculations will no longer be infinite, at any finite sample size, and the tail scores functions will be more nearly normal, justifying the normal approximations used in constructing the approximate distribution.

CHAPTER 5

Nonparametric Maximum Likelihood

We now return to the nonparametric maximum likelihood problem that was introduced in Section 1.6 of Chapter 1, and do the necessary theory to prove the results given there.

The problem is to maximize the mixture likelihood

$$L(Q) = \prod_{i=1}^{n} L_i(Q) = \prod_{j=1}^{D} \left[\int L_j(\phi)\, dQ(\phi) \right]^{n_j}. \tag{5.1}$$

Here $L_j(\phi)$ is the *likelihood kernel*, generally the one-component likelihood for a single observation, say y_j, and n_j is the number of times y_j was observed. The likelihood kernel may well depend on other auxiliary parameters and covariates, which will be held fixed in this discussion. As far as the maximization problem is concerned, the only critical assumption is that L_j is a nonnegative function of ϕ and that the number D is minimal among all such product representations. That is, the terms have been grouped to the maximal extent. In the multinomial setting, this can substantially reduce the number of terms in the product.

5.1. The optimization framework. The basic results concerning the nonparametric maximum likelihood estimator $\hat{Q}$ have already been outlined in Section 1.6. These results can be derived by putting the problem of likelihood maximization into the formal setting of numerical optimization theory. That is, we view it as a problem of the form: maximize an *objective function* $l(\mathbf{p})$ over the elements $\mathbf{p}$ of a set $\mathbf{P}$. If this is done properly, then the results follow readily from standard optimization results.

5.1.1. *Reformulating the problem.* The key to putting this problem into this framework is to examine (5.1) and recognize that the maximum depends directly on the possible values of the *mixture likelihood vector*

$$\mathbf{L}(Q) = (L_1(Q), L_2(Q), \ldots, L_D(Q))'.$$

We change our perspective on this problem from maximizing the likelihood over all latent distributions Q into the problem of determining which of the eligible classes of mixture likelihood vectors $\mathbf{L}(Q)$ gives the largest value to the likelihood.

We break the formulation into three steps.

STEP 1. Construct the *feasible region* of $\mathcal{R}^D$. It will be the set of all possible fitted values of the likelihood vector:

$$\mathcal{M} = \{\mathbf{L}(Q) = (L_1(Q), \ldots, L_D(Q))' \colon Q \text{ a probability measure}\}.$$

We have already seen sets of this type in Chapter 2.

STEP 2. Define the appropriate *objective function*, here (using the log likelihood)

$$l(\mathbf{p}) := \sum_{j=1}^{K} n_j \ln(p_j),$$

which we wish to maximize over all $\mathbf{p} \in \mathcal{M}$.

Suppose we have found that element of $\mathcal{M}$, say $\hat{\mathbf{L}}$, that maximizes this objective function.

If we were to solve the mixture problem using only these two steps, then there is one more step to carry out:

STEP 3. Solve for the maximum likelihood estimators $\hat{Q}$ by solving from the known $\hat{\mathbf{L}}$ for the latent distribution $\hat{Q}$ via the D equations

$$\mathbf{L}(\hat{Q}) = \hat{\mathbf{L}}.$$

It is instructive to compare this formulation of the problem to the normal theory least squares problem. In the latter one minimizes the objective function $\sum(y_i - \hat{y}_i)^2$ over the feasible set $\mathcal{F}$ that consists of all vectors $\hat{\mathbf{y}}$ of possible fitted values under the model. Corresponding to Step 3, the regression parameters $\hat{\beta}$ can therefrom be determined by solving, from $\hat{\mathbf{y}}$,

$$\hat{\mathbf{y}} = \mathbf{X}\hat{\beta}.$$

Much of the theory we derive here merely says that our optimization problem has many of the nice features of the linear regression problem. We examine the key structural features that give us this result.

5.1.2. *The feasible region.* The feasible region $\mathcal{M}$ has two key features that are of importance to us in the optimization problem. First, it is a convex set. As we shall see, this together with the concavity of our objective function, ensures that this is in a class of nice optimization problems.

The second key feature is that the convex feasible region can be expressed as the convex hull of a basic set Γ. In our case, if we define the *unicomponent likelihood vector* to be

$$\mathbf{L}_\phi = (L_1(\phi), \ldots, L_D(\phi))',$$

the mixture likelihood vector for the unicomponent model with parameter ϕ, then

$$\mathbf{L}(Q) = \int \mathbf{L}(\phi)\, dQ(\phi).$$

It follows that if we define the *unicomponent likelihood curve*

$$\Gamma = \mathbf{L}(\phi) \colon \phi \in \Omega\},$$

the desired mixture set is then

$$\mathcal{M} = \text{conv}(\Gamma).$$

The elements of Γ can be thought of as serving as the convex version of a basis, in that we can represent all eligible mixture vectors by convex combinations from this basic set.

We note the convex hull representation distinguishes this problem somewhat from the standard convex optimization problem, in which the convex region is expressed in terms of constraints that are satisfied by elements of the set.

We have already considered convex hull representations of a similar type in Chapter 2. In that chapter we considered the convex hulls of unicomponent density vectors $\mathbf{f}_\phi$, where

$$\sum_t f(t; \phi) = 1.$$

It follows that if the likelihood kernels are multinomial densities $f(t; \phi)$, and $n_t > 0$ for all t, then we can equate the likelihood vector $\mathbf{L}_\phi$ with the density vector $\mathbf{f}_\phi$. Otherwise, even in the multinomial model, the mixture likelihood vectors do not lie in the probability simplex, as we omit components for which $n_t = 0$.

When the likelihood is smoothly parameterized, then Γ is a curve. For example, suppose $f(x; \phi)$ is the Cauchy location density

$$\pi^{-1}[1 + (x - \phi)^2]^{-1}.$$

In Figure 5.1, we show the curve Γ for a pair of observations (y_1, y_2) that are separated by two units, such as $(-1, +1)$, so that the curve has the form

$$\Gamma = \{[1 + (1 - \phi)^2]^{-1}, [1 + (1 + \phi)^2]^{-1} : \phi \in \Omega\}.$$

(We have done the usual simplification of removing constant factors, here π^{-1}, from the likelihood.) The convex hull $\mathcal{M}$ of Γ includes the regions bounded by the dashed lines.

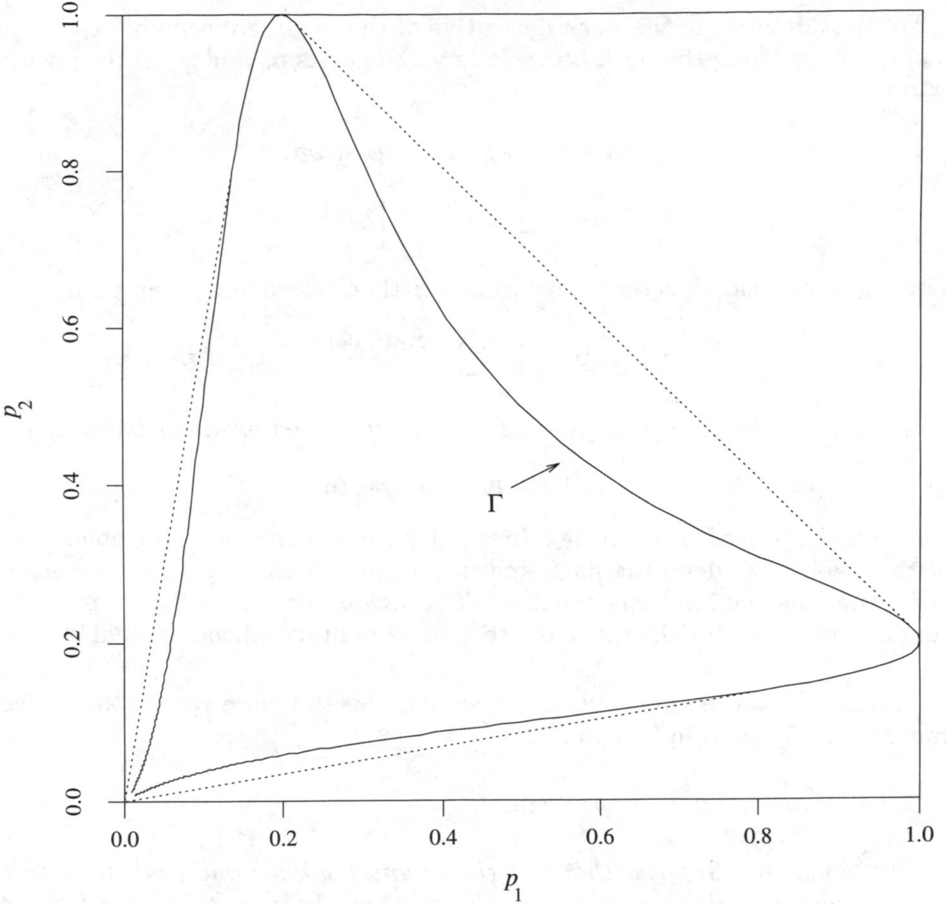

FIG. 5.1. *The unicomponent likelihood curve* Γ *for two Cauchy observations.*

It is not difficult to show that if the set Γ, is closed, then conv(Γ) is closed. This will be of some importance to the existence of the nonparametric MLE, so we offer some comments after stating the theorem.

5.1.3. *The objective function.* The second key feature of our reformulated optimization problem is the concavity of the objective function. Since we are working in a convex set with a convex hull representation, it is natural to examine the properties of the objective function in terms of its behavior along *paths*

$$\mathbf{p}_\varepsilon = [1 - \varepsilon]\mathbf{p}_0 + \varepsilon \mathbf{p}_1$$

between pairs $(\mathbf{p}_0, \mathbf{p}_1)$ of elements of the convex set. Restricted to this path, the objective function can be viewed a function of the parameter ε.

We first determine the path derivative of the log likelihood objective function as we go along the path between any two points $\mathbf{p}_0$ and $\mathbf{p}_1$ in the positive orthant:

$$d_1(\mathbf{p}_0, \mathbf{p}_1) := \frac{d}{d\varepsilon} l([1-\varepsilon]\mathbf{p}_0 + \varepsilon\mathbf{p}_1)\Big|_{\varepsilon=0} = \sum_i \left[\frac{p_{2i}}{p_{1i}} - 1\right] n_i. \tag{5.2}$$

The log likelihood objective function is strictly concave along any path:

$$\frac{d^2}{d\varepsilon^2} l([1-\varepsilon]\mathbf{p}_0 + \varepsilon\mathbf{p}_1) = -\sum \frac{(p_{1i} - p_{0i})^2}{p_{\varepsilon i}^2} < 0 \quad \text{if } \mathbf{p}_0 \neq \mathbf{p}_1.$$

It follows that for any $\mathbf{p}_0 \neq \mathbf{p}_1$, we have the *likelihood-gradient inequality*

$$l(\mathbf{p}_1) \leq l(\mathbf{p}_0) + d_1(\mathbf{p}_0, \mathbf{p}_1). \tag{5.3}$$

(This can be proved by creating a first order Taylor expansion in ε about $\varepsilon = 0$ of the likelihood along the path and using the second derivative property to show that the remainder is negative.) This inequality will suffice to prove our fundamental results about the mixture maximum likelihood estimator.

5.2. Basic theorems. We are now ready for the main results, here given more formally than in Chapter 1.

5.2.1. *Existence and support size.*

THEOREM 18. *Suppose that Γ is closed and bounded and that $\mathcal{M}$ contains at least one point with positive likelihood. Then there exists unique $\hat{\mathbf{L}} \in \partial\mathcal{M}$, the boundary of $\mathcal{M}$, such that $\hat{\mathbf{L}}$ maximizes $l(\mathbf{p})$ over $\mathcal{M}$.*

The statement is a slight correction of Lindsay (1983a), which failed to state that if *no point* in $\mathcal{M}$ has positive likelihood, then the uniqueness of the maximum must fail, because then all elements have likelihood zero. This theorem corresponds to Parts 1 and 4 of the mixture NPMLE theorem described in Chapter 1.

The proof is an elementary application of a fundamental result from optimization. We invite the interested reader to consult a general book on convex optimization, such as Roberts and Varberg (1973), to gain further perspective on the following description.

The objective function l is strictly concave on the positive orthant. In particular, this means that the upper sets

$$\mathbf{U}_c = \{\mathbf{p} : l(\mathbf{p}) \geq c\}$$

are closed convex sets. Since Γ is closed and bounded, so is $\mathcal{M}$, and therefore the likelihood objective function l takes on some maximum value at some

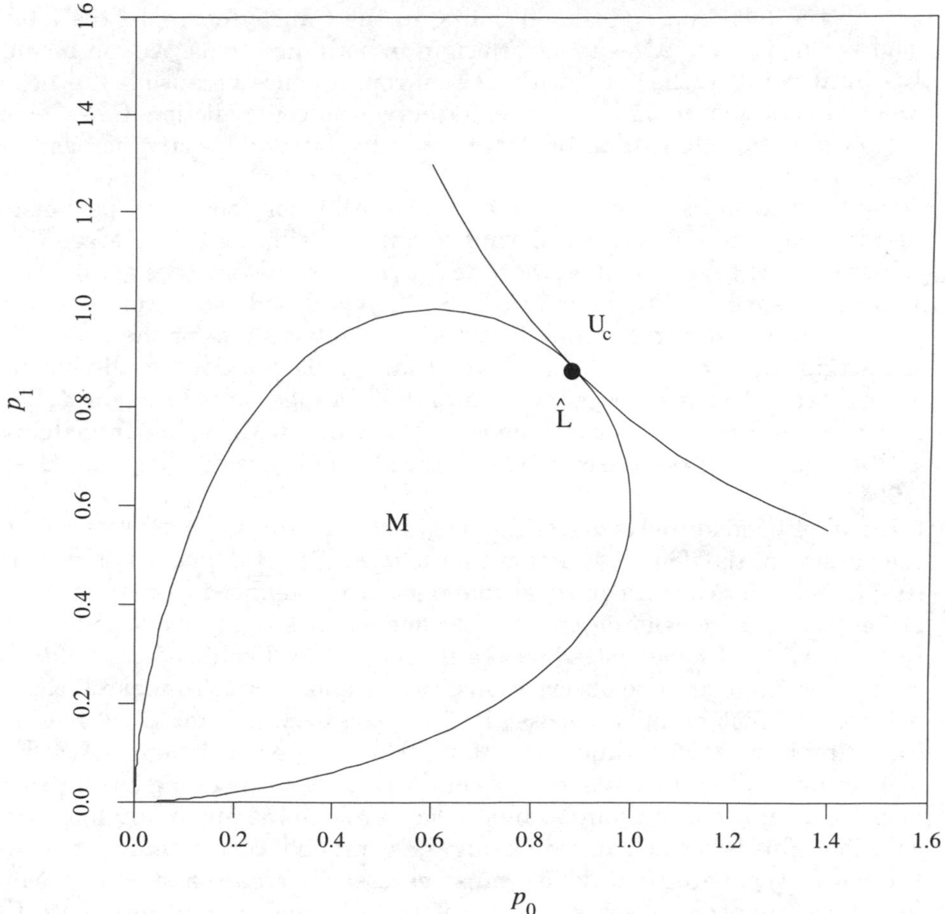

FIG. 5.2. *The geometry of the likelihood maximization problem, showing the unique solution.*

point with strictly positive likelihood. Geometrically, that uniqueness can be seen to correspond to the unique contact point between the upper set $\mathbf{U} = \{\mathbf{p}: l(\mathbf{p}) \geq l(\hat{\mathbf{L}})\}$ and $\mathscr{M}$. (See Figure 5.2.)

5.2.2. *Closed and bounded?* Before proceeding, we wish to address one question about the preceding result that sometimes arises in application. Boundedness of the curve Γ is essential, because if the likelihood vectors have unbounded components, then one can construct unbounded likelihoods. However, the requirement that the set Γ be closed is more of a technical requirement to make the theory simple. I have not found a case where this is a significant issue. To illustrate, we consider some examples.

- There will be cases, such as the mixture of Cauchy densities above, where the parameter ϕ has the range $(-\infty, +\infty)$. To ensure closure of Γ, we must include the left- and right-hand limits; in the Cauchy example, the likelihood vector $\mathbf{L}_\phi$ converges to $\mathbf{0}$, the origin in both directions. We can include this limit point in the set Γ without real consequence because it can never appear in the maximizing mixture. (Otherwise a contradiction arises, since one could eliminate it from the latent distribution and strictly increase the likelihood.)

- Consider next a distribution such as $\text{Bin}(n, p)$, whose boundary parameter values $p = 0$ and $p = 1$ correspond to true distributions. These are limit points of Γ, so that even if we were to set the parameter space as $(0, 1)$, we must necessarily include them in Γ. Since $p = 0$ and $p = 1$ correspond to $\phi = \pm\infty$ in the natural parameterization, we must allow for the possibility of putting mass at ∞ in our estimated latent distribution for the natural parameter unless we prespecify a finite closed range, say $[L, U]$ for ϕ. This comment applies to many contingency table models with log linear modeling, and is relevant in the Rasch model discussion of Lindsay, Clogg and Grego (1991).
- If the likelihood kernel is discontinuous in ϕ, then the set $\mathscr{M}$ may depend on the version of the density function that is used. For example, if f is uniform $(0, \phi)$, then there are two natural versions of the likelihood $L(\phi; x)$ at $\phi = x$, either $1/x$ or 0, depending on whether one chooses right or left continuity: In Figure 5.3 we have plotted such a unicomponent likelihood curve for the case when there are two observations, $x_1 = 1$ and $x_2 = 4$. To make Γ closed, we need to include all the possible limit points, which for $\phi = 4$ means including both $(0.25, 0)$ and $(0.25, 0.25)$. However, even though the closure of Γ appears then to contain two points $\mathbf{L}(\phi)$ corresponding to the same value of ϕ, only one of them is able to play a role in the maximum likelihood solution. This is because mixing using the point on Γ corresponding to using the value $1/x$, here $(0.25, 0.25)$, must necessarily create a strictly greater likelihood than the other value, here $(0.25, 0)$, and so will eliminate the other from being in the final mixture. [*Exercise.*] This remark can clearly be applied to any similar univariate parameter likelihood where the individual components each display a distinct finite set of discontinuities—while in theory we would need to include both right and left limits to apply the theorem, the maximum likelihood estimator will only use the limit point with the larger component values.

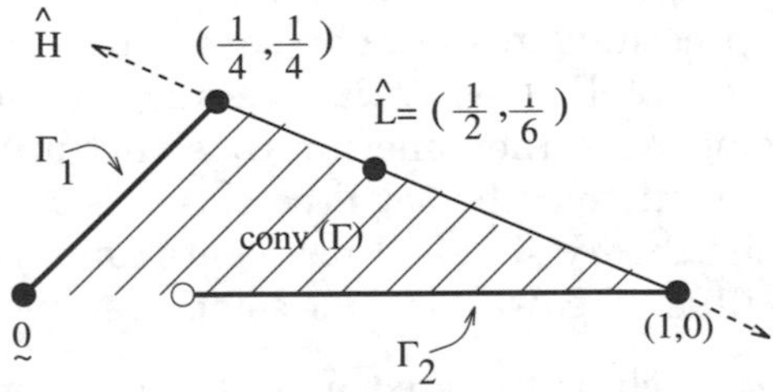

FIG. 5.3. *The uniform likelihood curve, together with its limit points.*

The moral of this story is that, for many problems, as long as one is careful about defining the parameter space to be a closed set and the likelihood vector components are defined to be maximal at discontinuities, there will exist a unique maximum likelihood estimator $\hat{\mathbf{L}}$, even if Γ is not technically closed.

5.2.3. *Gradient characterization.* We now turn to the second part of the mixture NPMLE theorem, the gradient characterization. Recall from Section 1.6 of Chapter 1 the gradient function

$$D_Q(\phi) = \sum_i n_i \left[\frac{L_i(\phi)}{L_i(Q)} - 1 \right].$$

As derived there, this is the path derivative of the log likelihood $\ln(L(Q))$ for the one parameter mixture

$$(1-\pi)Q + \pi\Delta_\phi,$$

evaluated at $\pi = 0$.

We can relate this to our formal optimization theory as follows. From this we see that

$$d_1(\mathbf{L}(Q), \mathbf{L}(\phi)) = D_Q(\phi).$$

That is, the gradient function equals the first derivative of the objective function l along the path from the current fitted vector toward a basic vector. (That the derivatives match up in this way is a consequence of the linear structure of the mixture model). Put into this form, we are then in the setting for classical optimization theory. We can record the basic result in terms of a theorem:

THEOREM 19. *The following three statements are equivalent:*

1. $\hat{Q}$ *maximizes* $l(Q)$.
2. $\hat{Q}$ *minimizes* $\sup_\phi D_Q(\phi)$.
3. $\sup_\phi \{D_{\hat{Q}}(\phi)\} = 0$.

PROOF. If we let $\hat{\mathbf{L}}$ play the role of $\mathbf{p}_0$ in the likelihood-gradient inequality, we see that for $\hat{Q}$ to maximize the likelihood, it suffices that

$$d_1(\hat{\mathbf{L}}, L(Q)) \leq 0,$$

for all Q. Therefore, it is sufficient that

$$d_1(\hat{\mathbf{L}}, L(\phi)) \leq 0,$$

for all ϕ; hence statement 3 implies 1. Item 1 implies 3 because if the gradient is anywhere positive, we can necessarily increase the likelihood along that path. Finally, $\hat{Q}$ must minimize the sup gradient at the value 0, because if there were Q_1 with supremum less than zero, we could use the likelihood-gradient inequality (5.3) to show Q_1 has greater likelihood than $\hat{Q}$. □

5.2.4. *Properties of the support set.* This completes the proof of the second part of the theorem. Now, the third part:

THEOREM 20. *The support of any maximum likelihood estimator $\hat{Q}$ lies in the set*

$$\{\phi: D_{\hat{Q}}(\phi) = 0\}.$$

PROOF. Consider the one parameter family of mixtures

$$Q_\varepsilon := (1-\varepsilon)\hat{Q} + \varepsilon\Delta_\phi.$$

If ϕ is a support point of $\hat{Q}$, then Q_ε continues to be a true probability measure for some negative values of ε. It follows that the maximum value is taken on an interior point of the allowable range of ε. This implies that the derivative of the likelihood along the one parameter path equals zero at this point; however, this derivative is just the gradient function $D_{\hat{Q}}(\phi)$. □

These results have very simple geometric interpretations. For any candidate mixture likelihood vector $\mathbf{L}$ in $\mathcal{M}$, the gradient function determines a hyperplane

$$\mathcal{H} := \{\mathbf{p}: d_1(\mathbf{L}, \mathbf{p}) = 0\}$$

that contains the point $\mathbf{L}$. If $\mathbf{L}$ is indeed $\hat{\mathbf{L}}$, the maximum likelihood point, then this hyperplane $\mathcal{H}$ is a support hyperplane to the set $\mathcal{M}$ and separates that convex set from the convex upper set of the log likelihood objective function $\mathbf{U} = \{\mathbf{p}: l(\mathbf{p}) \geq l(\hat{\mathbf{L}})\}$. Lying exactly in the support hyperplane are all the support vectors $\mathbf{L}_{\phi_i}$. See Figure 5.4. This interpretation allows us to apply Carathéodory's theorem (Section 2.3.4) to characterize the existence of a discrete latent distribution that maximizes the likelihood:

THEOREM 21. *The solution $\hat{\mathbf{L}}$ can be represented as $\mathbf{L}(\hat{Q})$, where $\hat{Q}$ has no more than D points of support.*

We note that if we are in the setting of the multinomial exponential family of Chapter 2, we can employ the superior bounds on the mixture representations that were given there, with a bound of roughly $D/2$. (See the discussion of index in Chapter 2 for more precise descriptions.) However, we emphatically note that it was *absolutely critical* that the mixture likelihood vectors lie in the probability simplex for this reduction to take place. For example, in a normal mixture model with fixed σ^2, one can construct sets of data for which the bound D is attained simply by spreading the observations widely apart.

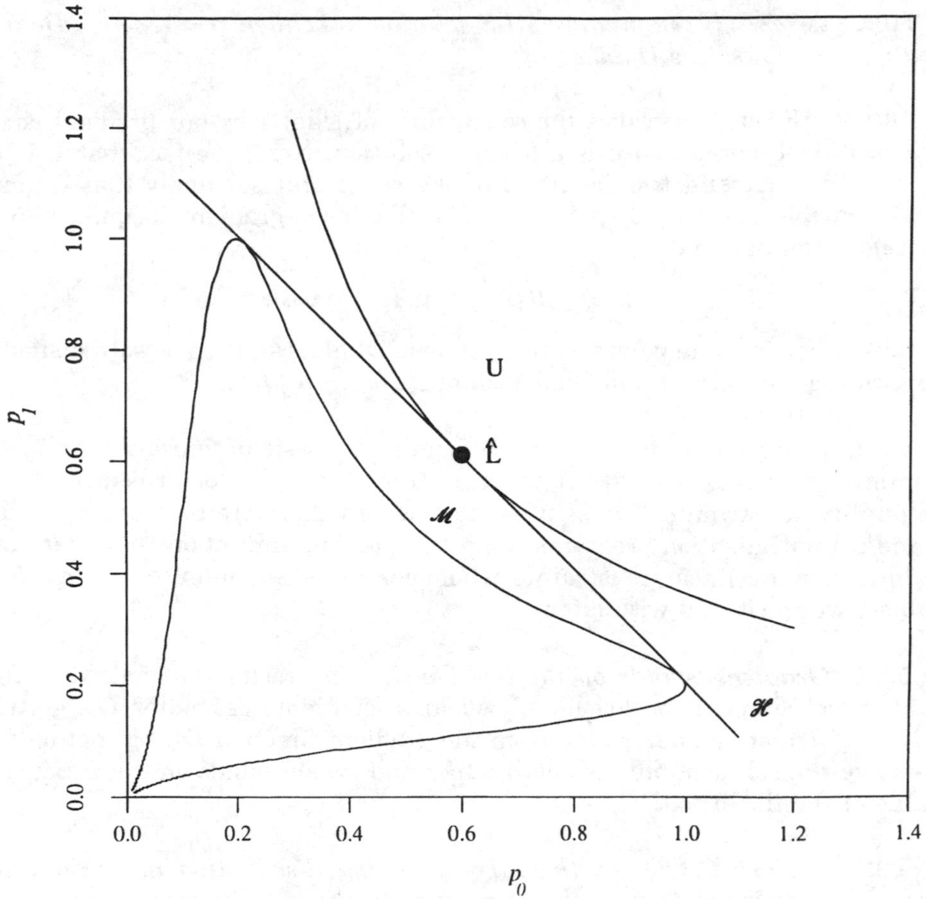

FIG. 5.4. *The support hyperplane of the maximum likelihood point.*

5.3. Further implications of the theorems.

5.3.1. *Duality theorem.* We specify two versions of the dual problem, with the understanding that they are equivalent, but the first one arises naturally in the context of the problem, whereas the second is in the form of a classical optimization problem:

DUAL 1. Minimize $l(\mathbf{p})$ subject to the constraints $\mathbf{p} \geq \mathbf{0}$, and $d_1(\mathbf{p}, \mathbf{L}(\phi)) \leq 0$, for all $\phi \in \Omega$.

DUAL 2. Maximize $l(\mathbf{w})$ subject to the constraints $\mathbf{w} \geq \mathbf{0}$ and $\sum w_i L_i(\phi) \leq 0$, for all $\phi \in \Omega$.

THEOREM 22. *If the mixture MLE solution is* $\hat{\mathbf{L}}$, *then* $\mathbf{p} = \hat{\mathbf{L}}$ *solves Dual* 1 *and* $\hat{w}_i = n_i/\hat{L}_i$ *solves Dual* 2.

PROOF. First, $\hat{\mathbf{L}}$ satisfies the constraints of Dual 1 by our gradient characterization theorem, so it is a feasible solution. Next, the fact that $\hat{\mathbf{L}}$ is in the mixture set and the definition of the constraint set imply that for any other feasible point $\mathbf{p}$, $d_1(\mathbf{p}, \hat{\mathbf{L}}) \le 0$. The likelihood-gradient inequality (5.3) therefore implies that

$$l(\hat{\mathbf{L}}) \le l(\mathbf{p}) + d(\mathbf{p}, \hat{\mathbf{L}}) \le l(\mathbf{p}),$$

for any other feasible $\mathbf{p}$, as was to be shown. Dual 2 can then be solved simply by a change in variables in Dual 1, namely, $w_i := n_i/p_i$. □

We now note that Dual 2 has the form of a classic optimization problem: minimize a convex objective function $(-l)$ over a region described by linear inequality constraints. The solution to this can therefore be found by using standard optimization programs, with the possible limitation that there are *infinitely* many linear constraints whenever there are infinitely many ϕ, a subject we must deal with later.

5.3.2. *Gradient bounds on the likelihood.* The result that follows shows that if we compute the log likelihood at a candidate estimator Q_c, getting $l(\mathbf{L}_{Q_c})$, then we can determine from the gradient function $D_{Q_c}(\phi)$ not only if we have the solution, but also both upper and lower bounds on the maximum value of the likelihood.

THEOREM 23. *Let* Q_c *be the current mixing distribution in an iterative algorithm designed to find the maximum likelihood estimator. Define* $\delta = \sup_\phi D_{Q_c}(\phi)$. *Then*

$$A(\delta) \le l(\hat{\mathbf{L}}) - l(L_{Q_c}) \le B(\delta) \le \delta,$$

where $B(\delta) := n \ln(1 + \delta/n)$, $A(\delta) := B(\delta) - n^* \ln[1 + \delta/n^*]$ *and* $n^* = n - \min_k\{n_i\}$.

PROOF. For the lower bound, see the not-so-simple argument in Lindsay (1983b). For the upper bound, we can use Dual 2. The point

$$\left[\frac{n}{n+\delta}\right]\mathbf{L}_{Q_c}$$

meets the linear constraints of the optimization problem, so it is a feasible solution. It follows that

$$l\left(\left[\frac{n}{n+\delta}\right]\mathbf{L}_{Q_c}\right) \le l(\hat{\mathbf{w}}),$$

which gives the desired upper bound. □

In any algorithm, it is useful to have a way to determine how much more computation might be needed to converge to the solution. The preceding result shows that the maximum of the gradient function can be used for this. We will consider convergence criteria for algorithms further in the next chapter.

5.3.3. *Link to m-component methods.* The preceding results enable us to contrast the properties of the *global* maximum likelihood estimator of the latent distribution $\hat{Q}$, and estimators formed by maximizing the likelihood, or solving the likelihood equations, when the latent distribution is restricted to having a fixed number of support points. In this section, we extend the results of Section 3.3.

Let $\hat{Q}_m$ be a latent distribution, such as described in Chapter 3, that maximizes the m-point mixture likelihood. Earlier we derived part of the relationship between the gradient function and the EM algorithm that is often used to find $\hat{Q}_m$. That is, we noted that the EM algorithm for the weights can be written as

$$\hat{\pi}_{j,\text{new}} = \hat{\pi}_{j,\text{old}}[1 + n^{-1}D_Q(\phi)],$$

so that the weights increase or decrease according to the sign and magnitude of the gradient function. In addition, it is easy to check that the new support points move left or right from the old ones in agreement with the direction of the greater gradient. [That is, if $D'_Q(\phi)$ is positive at a support point ϕ, then the EM algorithm puts the new point to the right.]

Next, the following basic results are from Lindsay (1981):

THEOREM 24. *Suppose ϕ^* is a support point of $\hat{Q}_m$. If the gradient function is twice differentiable in ϕ at ϕ^*, then:*

1. $D_{\hat{Q}_m}(\phi^*) = 0$.
2. $D'_{\hat{Q}_m}(\phi^*) = 0$.
3. $D''_{\hat{Q}_m}(\phi^*) \le \sum n_k[(L_k(\phi^*)/(L_k(\tilde{Q}_m))]^2$.

PROOF. [*Exercise.*] These arise in a straightforward way through manipulation of the likelihood equations and the formula for the gradient. □

The global MLE is, of course, an m-point MLE as well, for some m, so that it satisfies all three of the above properties. However, it satisfies a stronger property than statement 3 because each support point is at a local maximum of the gradient and so

$$D''_{\hat{Q}}(\phi^*) \le 0.$$

5.3.4. *Moment and support point properties.* We have already mentioned some properties of the nonparametric MLE in Section 5.2.4. We now mention a few further results that can be found in Lindsay (1981). One of the more important results is relevant in the search of the parameter space Ω for potential support points.

PROPOSITION 25. *Suppose that the parameter ϕ is real-valued and that for every i, the likelihood kernel $L_i(\phi)$ is unimodal in ϕ, with unique mode at $\tilde{\phi}_i$. Then all the support points of $\hat{Q}$ lie in the interval*

$$[\min_i \tilde{\phi}_i, \max_i \tilde{\phi}_i].$$

PROOF. [*Exercise.*] Show that the gradient function is increasing whenever $\phi \leq \min_i \tilde{\phi}_i$ and decreasing for $\phi \geq \max_i \tilde{\phi}_i$. Since the latent support points are in the set of maxima to the gradient, this proves the result. □

The next result relates to the dispersion score v_2 of Neyman and Scott that was introduced in Chapter 4. We recall from Section 4.1 that if the one-component solution is also the NPMLE, then it is necessarily true that

$$\sum v_2(\hat{\phi}, x_i) \leq 0,$$

because this score is also $D_1''(\hat{\phi})$ and we would otherwise have a local failure of the gradient inequality. We can extend this result to any of the support points of $\hat{Q}$, say ϕ^*, by recalling the second derivative inequality, $D_{\hat{Q}}''(\phi^*) \leq 0$. With some further manipulation, we can relate this again to the dispersion score in that it is equivalent to requiring that

$$\sum E[v_2(\Phi, x_i)h(\Phi) \mid X = x_i;\ \hat{Q}] \leq 0,$$

for every nonnegative function $h(\phi)$. In the one parameter exponential family, (2.2), one can use the last equation to show that the sample variance is always smaller than the estimated variance under the model:

$$n^{-1}\sum(x_i - \bar{x})^2 \leq \mathrm{Var}(X; \hat{Q}).$$

One can strengthen this dispersion result—that the fitted model is biased in the direction of overdispersion—even further by using the full force of the gradient inequality. The following is a challenging exercise: Let $\xi_1, \ldots, \xi_m$ be the support set for $\hat{Q}$ in a one parameter exponential family, in the natural parameter. For every value of t such that $t+\xi_i$ is in the parameter space for all i, the following inequality of sample and model moment generating functions holds:

$$n^{-1}\sum e^{tx_i} \leq \int e^{tx}\, dF(x; \hat{Q}).$$

This result is not in Lindsay (1981), but can be proved by using the gradient inequality and the fact that the exponential family density has exponential form. Differentiating twice with respect to t leads to the variance inequality above.

5.4. Applications. At this point we offer some simple examples that may help elucidate these mathematical structures and methods.

5.4.1. *A binomial mixture.* First, we have some plots that illustrate how one can use the gradient function. In Figure 5.5 we have shown a sequence of plots for a particular sequential optimization scheme. The data are the sibship data of Chapter 2, and we use the binomial mixture model discussed there. In the algorithm that was used, the nonparametric mixture estimator was found by first finding the best *one* support point model, then the best *two* and so on, until we find that we can no longer increase the likelihood by adding support points; that is, the gradient inequality is satisfied. One can use the EM algorithm with a fixed number of support points, iterate until convergence and then check the gradient function. If it is greater than zero at some point ϕ, then one can add that support point to the rest in a way that increases the likelihood and then return to the EM, but with one larger support size.

The first plot shows the best two support point model. In accordance with the above discussion, we see that the two support points show up as zeros and

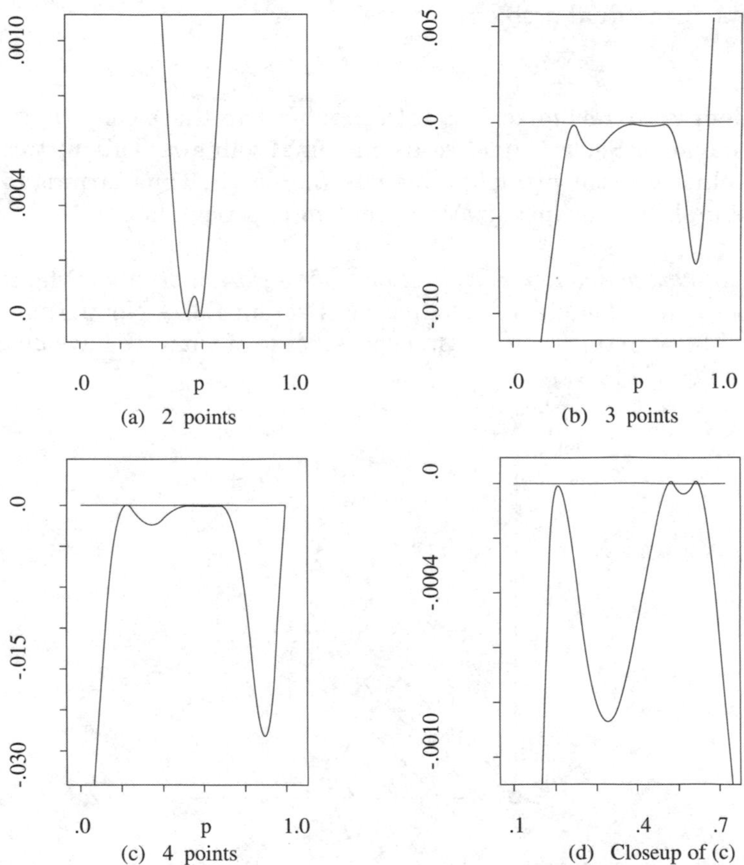

FIG. 5.5. *The sequence of gradient plots for the sibship data.*

the derivative of the gradient is zero there. However, the gradient inequality is certainly violated, and indeed the support points are local minima to the gradient. Next, the three point fit also violates the gradient inequality and we can see that the likelihood can be increased by adding mass near $p = 1$. When this is done, we achieve the gradient inequality, as can be seen in the lower two frames.

5.4.2. *Empirical CDF.* In Figure 5.6 we show the rather simple geometry that arises for the nonparametric distribution function problem of Section 1.7.1. In this case, the likelihood kernel was $L_i(\phi) = \mathscr{I}[\phi = x_i]$. Thus for a sample of size two, the likelihood "curve" Γ consists of three points only: $(1, 0)$ at $\phi = x_1$, $(0, 1)$ at $\phi = x_2$ and $(0, 0)$ for any other value of ϕ. As noted earlier, the nonparametric maximum likelihood estimator corresponds to the sample proportions at the two observed values, a point on the simplex determined by the convex hull of these three points.

We can use this problem to illustrate the dual problem approach. Here the constraints of the dual problem become

$$\sum w_i L_i(\phi) \leq n \qquad \forall \phi \Leftrightarrow w_i \leq n.$$

The problem is to maximize $\sum n_i \ln(w_i)$ subject to these constraints. The solution is obvious: Set w_i equal to its maximal value n. This in turn implies that the solution to the primal problem is $\hat{L}_i = n_i/n$. Thus turning to the dual problem simplifies the optimization problem to a triviality.

5.4.3. *Known component distributions.* We return briefly to the case of the known component densities (introduced in Section 1.3.1). Nonparametric maximum likelihood in this case is just a special case of what we have described, in

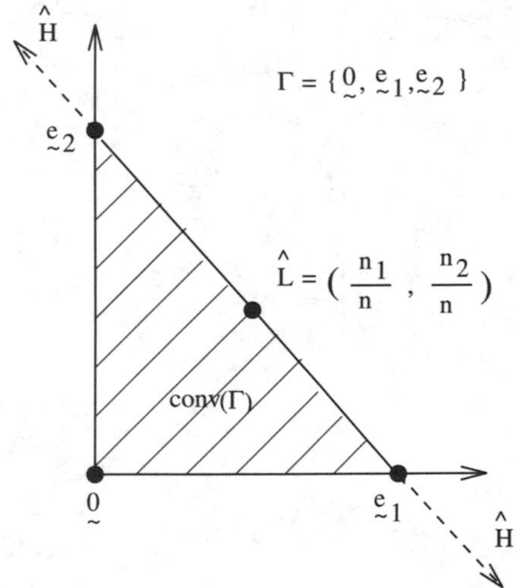

FIG. 5.6. *The likelihood curve for the nonparametric distribution problem.*

which the latent variable ϕ is just the index of the known component density. Roeder, Devlin and Lindsay (1989) give a relatively complete description of the implications of the theorem in this setting. In addition, see Vardi and Lee (1993), where one can find references to literature in areas other than mixture models in which pieces of the nonparametric maximum likelihood theory have been developed.

5.4.4. *The multinomial case.* We recall the multinomial geometry discussed in Section 2.3. One nice feature of the multinomial problem is that one can picture the behavior of the nonparametric maximum likelihood estimator $\hat{\mathbf{L}}$ as a function of the sample frequencies vector, denoted by $\mathbf{d}$ in Chapter 4. For example, it is clear that if $\mathbf{d}$ falls in the set of eligible mixture density vectors, then $\hat{\mathbf{L}} = \mathbf{d}$ (since $\mathbf{d}$ always globally maximizes the likelihood objective function l over the entire probability simplex and by assumption is also a mixture vector). This property was used in Lindsay, Clogg and Grego (1991) to show that the nonparametric mixture approach was equivalent to a conditional approach for estimating certain auxiliary regression parameters.

If $\mathbf{d}$ is not in the set of mixture vectors, then one can partition the probability simplex into regions corresponding to the number of components in the maximum likelihood solution. The reader is invited to consider how this might be done using the examples of Chapter 2.

5.5. Uniqueness and support size results. The final problem we address is the uniqueness of the estimator of the latent distribution, saved for last due to its technical difficulty. We will here just describe the basic issues and refer the reader desirous of more details to other sources.

Thus, to date, we have shown the uniqueness of the estimated mixture likelihood vector $\hat{\mathbf{L}}$. Can we infer from this the uniqueness of the latent distribution estimator $\hat{Q}$? That is, can we infer from $\hat{\mathbf{L}}$ the latent distribution itself by solving for $\hat{Q}$ in

$$\hat{\mathbf{L}} = \int \mathbf{L}(\phi)\, dQ(\phi).$$

From a geometric point of view, this may seem unlikely because the interior of the convex hull of a curve, such as $\mathcal{M} = \text{conv}(\Gamma)$, generally has infinitely many representations in terms of elements of its generating set Γ. (We saw this in Chapter 2.) What saves the day for us is that the solution $\hat{\mathbf{L}}$ is on the boundary of $\mathcal{M}$. We describe a strategy for proving uniqueness that works for many important examples.

5.5.1. *The strategy.* We first note that the gradient function at the maximum likelihood estimator is completely determined by $\hat{\mathbf{L}}$ and is not related to the choice of $\hat{Q}$. Thus from $\hat{\mathbf{L}}$ we can unambiguously determine the full set of points $\{\phi_1, \ldots, \phi_r\}$ that satisfy

$$D_{\hat{Q}}(\phi) = 0 \tag{5.4}$$

and so are candidates to be in the support set of $\hat{Q}$.

TASK 1. Show that (5.4) has at most $r \leq D$ solutions, subject to $D_{\hat{Q}}(\phi) \leq 0$; that is, the solutions must also be local maxima. The proof of this property must depend on the gradient function having a polynomial type structure that bounds the number of solutions to such equations. [Although this is related to the ideas of Chebyshev systems (Chapter 2), those ideas are not strong enough to work here.]

If Task 1 is completed, the set of possible support points is fixed at $\{\phi_1, \ldots, \phi_r\}$ and we proceed to:

TASK 2. We must now see if the weights are uniquely determined by the equations

$$\sum \pi_j \mathbf{L}(\phi_j) = \hat{\mathbf{L}}.$$

This is just a set of linear equations in π_j, so it suffices to show that every set of D unicomponent likelihood vectors $\{\mathbf{L}(\phi_j)\}$ is linearly independent. This is simpler than Task 1, in that we can apply the ideas of Chebyshev systems directly to obtain this result, as discussed in Chapter 2.

5.5.2. *A geometric approach to Task 1.* The first proofs of uniqueness used special properties of the likelihood kernel involved to complete Task 1 [for the Poisson model, see Simar (1976); for the exponential, see Jewell (1982)]. In essence, in these cases, the gradient could be written as an *exponential polynomial*, and certain long known bounds on the number of zeros to these polynomials could be used [Pólya and Szegö (1925)]. However, this was a piecemeal approach to the problem, when clearly there were more general truths at work.

In Lindsay (1983a), the problem was attacked from a geometric point of view for a unicomponent exponential family and it was shown that the number of zeros to the gradient function could be bounded by considering the geometric structure of the unicomponent likelihood curve. In effect, there was a way to consider the complexity of the curve Γ that gave a bound on the number of support points; the bound depended on the number of zeros to a certain polynomial, but was always less than D.

These results are easy to visualize for the case of two observations. In Figure 5.7 we have shown how, in the case of the normal mixture, with scale parameter 1, the shape of the likelihood curve $\mathbf{L}(\phi)$ depends very strongly on the distance between the two observations. If the observations are 1 unit apart, say $x_1 = -0.5$, $x_2 = 0.5$, we get the "balloon" shape of $\mathbf{\Gamma}_1$, enclosing a convex region. It is clear that the NPML estimator has one support point. If the observations are 3 units apart, shown as $\mathbf{\Gamma}_3$, the curve has a substantial indentation and the corresponding latent distribution estimator has two support points. The boundary case occurs when the observations are exactly 2 units apart, shown as $\mathbf{\Gamma}_2$. In the two-dimensional case, these results can be obtained by analyzing the sign of the curvature of the likelihood curve. Unfortunately, these results were very difficult to obtain in higher dimensions, and hard to generalize outside the exponential family.

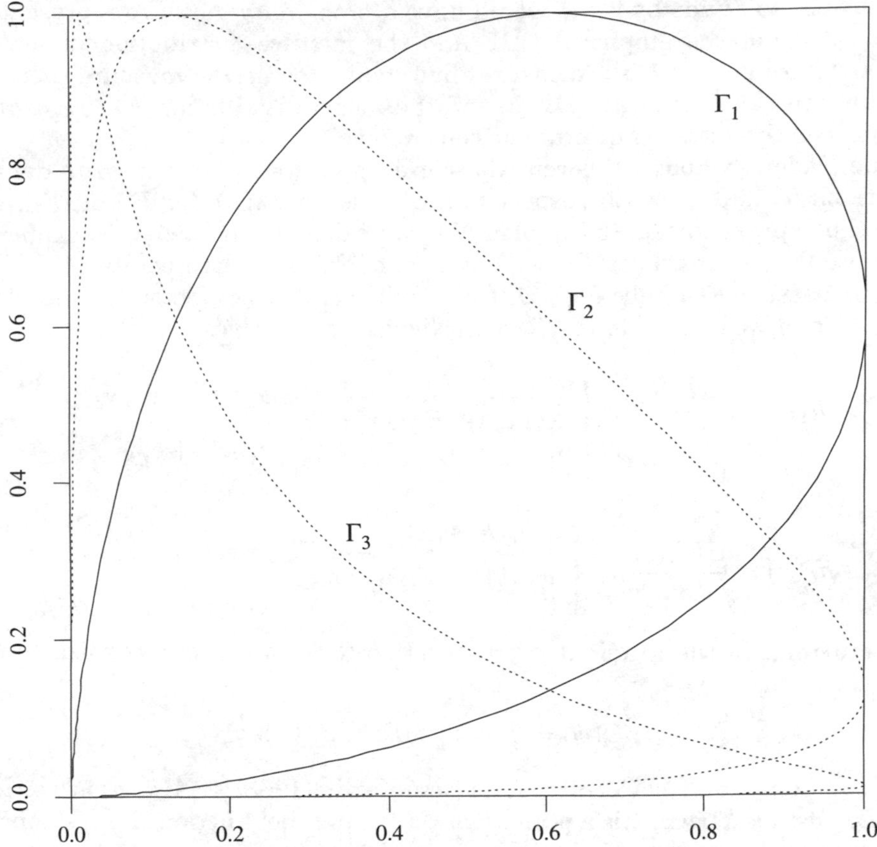

FIG. 5.7. *Three different unicomponent likelihood curves for the normal model.*

5.5.3. *A gradient function representation.* Therefore, we examine a great simplification of the proof. Since it relies on a powerful result from the theory of totally positive kernels, we merely illustrate the general method of attack here without pretending to be complete. For more, see Lindsay and Roeder (1993).

The key to the result is deriving a representation of the gradient function of the form

$$n^{-1}D_Q(\phi) = \int \left[\frac{f(x;\phi)}{f(x;Q)}\right] [h_1(x) - h_0(x)]\, d\lambda(x), \tag{5.5}$$

which we now construct. We start with the easy-to-prove representation

$$n^{-1}D_Q(\phi) = \int \left[\frac{f(x;\phi)}{f(x;Q)}\right] d[\hat{F}(x) - F_Q(x)]$$

Next, define the positive measure λ by the relationship

$$d\lambda(x) = d[\hat{F} + F_Q](x).$$

(If we were to divide by 2, we would have a probability measure representing the mixture of the empirical CDF and the mixture distribution under Q.) The key here is that λ is a measure that dominates (in the measure-theoretic sense) both the empirical CDF $\hat{F}$ and the model distribution F_Q, regardless of whether the latter is discrete or continuous.

The Radon–Nikodyn theorem therefore implies that we can write density functions h_1 and h_0, with respect to the measure $d\lambda(x)$, for $\hat{F}$ and F_Q, and that the representation (5.5) holds. We can calculate the densities explicitly: Suppose that $\hat{F}$ has mass $F(\{y_k\})$ at a set of observed data points $y_1, \ldots, y_K$. (In our case, most likely $K = D$, the number of distinct factors in the likelihood.) Then we can write the Radon–Nikodyn derivatives as

$$h_1(x) = \frac{d\hat{F}(x)}{d\lambda(x)} = \begin{cases} \dfrac{\hat{F}(\{x\})}{\hat{F}(\{x\}) + F_Q(\{x\})}, & x \in \{y_1, \ldots, y_K\}, \\ 0, & \text{otherwise,} \end{cases}$$

and

$$h_0(x) = \frac{dF_Q(x)}{d\lambda(x)} = \begin{cases} \dfrac{F_Q(\{x\})}{\hat{F}(\{x\}) + F_Q(\{x\})}, & x \in \{y_1, \ldots, y_K\}, \\ 1 & \text{otherwise.} \end{cases}$$

Returning to the gradient representation (5.5), we next recognize that it has the form

$$C(\phi) = \int A(x;\phi)B(x)\, d\lambda(x).$$

We next apply some powerful results from Karlin (1968). If $A(x;\phi)$ is a *strictly totally positive kernel*, λ is a positive measure and the function $B(x)$ is nonzero and has no more than M sign changes (relative to the measure λ), then the *variation diminishing property* of the totally positive kernel implies that $C(\phi)$ has no more than M sign changes unless it is identically zero.

Applying this to our case, we see that the difference in densities $B(x) = h_1(x) - h_0(x)$ can have at most $2K$ sign changes, one to each side of a data point. (It is negative between observations, but possibly positive at each observation.) It follows that if the family of densities is based on a totally positive kernel, then either the gradient will be identically zero or it will have at most K local maxima.

Lindsay and Roeder (1993) used this fact to show that in the exponential family, regardless of whether it is discrete or continuous, then either the latent estimator $\hat{Q}$ is unique *or* the gradient function is identically zero at the maximum likelihood solution and the latent estimator is nonunique. The latter can happen only with discrete f for which some mixtures are nonidentifiable, as in Chapter 2.

CHAPTER 6

Computation: The NPMLE

We now consider the computational issues involved in calculating the NPMLE. We will focus on the simplified version of the problem in which there are no auxiliary parameters. The standard strategy to incorporate auxiliary parameters is to alternate between an algorithm for the latent parameters and one for the auxiliary parameters; this section describes only the latent parameter phase of that operation.

We will start with an overview of the algorithmic strategies available, deferring the details to the works of others.

After the overview, we wish to address an important issue that has seen little attention. In most problems, one cannot hope to compute the NPMLE exactly because there is no finite time algorithm that will attain the solution. Thus one must devise strategies that ensure that the computations have gone far enough to give desired statistical accuracy, but have not gone needlessly far. We will offer one strategy for this.

At this point, there is a limited supply of software available for the nonparametric analysis. See Böhning, Schlattman and Lindsay (1992) for a description of C.A.MAN, Ezzet and Davies (1988) for a description of MIXTURE and DerSimonian (1986, 1990) for a published algorithm.

6.1. The convergence issue. We recall that the test for whether a *candidate* latent distribution Q is the nonparametric maximum likelihood estimator $\hat{Q}$ is to check whether the gradient inequality holds:

$$D_Q(\phi) \leq 0 \qquad \forall\, \phi \in \Omega.$$

Unfortunately, in a typical problem one has an iterative algorithm such that one cannot in a finite amount of time attain this inequality. There are two issues here.

First, there are often *infinitely* many inequalities to check, corresponding to all ϕ in Ω. We will later consider the implications of a simple solution to this problem where we assume that there is a chosen finite subset, say Ω_s, of s grid points ϕ_j, where the gradient will be checked. One of our points of interest becomes the appropriate choice for the elements of such a grid.

A more sophisticated approach, used by Lesperance and Kalbfleisch (1992), involves taking a basic grid and then doing further searches for gradient violations in the neighborhood of each grid point. An appropriate analysis of such a method requires rather more sophistication than our approach here. The key to a proper basic grid would be that there could not be regions between grid points that were left unsearched. One of the key points of our discussion here is that there are many situations in which we do not need this extra sophistication in programming. We can simply choose a grid to work on and iterate to a solution.

We note that, provided that one restricts the support points of Q to be also from the chosen finite grid, then the distribution Q that solves

$$D_Q(\phi) \leq 0 \qquad \forall\, \phi \in \Omega_s$$

is the nonparametric MLE among all latent distributions on the grid.

Unfortunately, even with such a restricted grid, one must still use an iterative algorithm that does not converge in a finite number of steps. We note in passing that when we restrict the maximization to a finite grid, we are exactly in the setting of known component densities, Section 3.2.1, where there is no known finite step algorithm.

One possible strategy at setting a stopping rule would therefore be to set a small positive tolerance on the values of the gradient function, such as γ, and set a stopping rule for the algorithmic process of the type

$$D_Q(\phi) \leq \gamma \qquad \forall\, \phi \in \Omega_s.$$

We return to this rule later, after reviewing some of the main algorithms that have been employed for this problem.

6.2. Using the EM. The simplest programming approach to finding the NPMLE is to follow the lead of Laird (1978), who suggested the use of the EM algorithm with a large number of support points. For example, one could use one support point for each data point, with locations suggested by the unicomponent maximum likelihood estimators. Provided one has the patience for a very slow algorithm and has a secondary strategy for monitoring the gradient function to ensure adequate convergence, this is a reasonable strategy. We note that even if there is not an explicit solution to the EM equations for the component parameters, one can always simply fix a grid of ϕ values and use the EM to solve for the unknown weights because this algorithm is always simply calculated.

See DerSimonian (1986, 1990) for details of an algorithm.

6.3. Gradient-based algorithms. For completeness, we here offer a description of some basic algorithmic issues and constructions. As more complete sources, we recommend Böhning, Schlattman and Lindsay (1992) and Böhning (1995).

6.3.1. *Design algorithms.* If we view the problem in the wide context as a convex optimization problem on a set defined by its convex hull representation, then the mixture problem is, of course, identical in structure to other problems. Of particular relevance to statisticians is the problem of the *optimal design of experiments*, where the *objective function* is one of several scalar measures of the overall informativeness of an experiment based on the resulting information matrix for the parameters. The *feasible region* is the space of information matrices that are allowed under the design constraints. The space of feasible design matrices can be shown to be the convex hull of a set of basic matrices determined by the selection of a single design point. See Silvey (1980) for an overview of this problem.

Of consequence here is that many of the algorithms used in mixtures can be used in design problems as well, including EM type algorithms, and of course, the converse is true. Thus a detailed study of the literature is necessarily quite extensive, and we will here only point to some of the mixture sources.

6.3.2. *Keeping track of the support points.* One technical issue that separates various algorithmic approaches concerns the kind of information that is stored and used at each iteration. The reason for this is that it is technically not necessary to keep track of the current latent distribution estimator Q_c at each step. If one instead merely updates the current fitted likelihood vector $\mathbf{L}_c$, uses this to construct the gradient function and therefrom an update of $\mathbf{L}_c$, say $\mathbf{L}_{c+1}$, then one may have everything one needs. Since there are a number of algorithms of this type, we need to say why.

It is because $\hat{Q}$ itself can in theory be reconstructed after convergence from $\hat{\mathbf{L}}$. Its support points ξ_j can be found from the final gradient function because they are in the set of points where the gradient reaches a local maximum of zero. The weights π_j, at least in theory, can then be determined after convergence by the relationship

$$\sum \pi_j \mathbf{L}(\xi_j) = \hat{\mathbf{L}}.$$

Such an algorithm might be contrasted with an EM algorithm that uses many support points—an approach recommended by some—because we must save the current parameter values for it so that they can be updated at the next step.

6.3.3. *Vertex direction and exchange methods.* In the following algorithmic methods, we will let Q_c be the current estimate of the latent distribution and we describe the construction of the next step.

The simplest algorithm for finding the NPMLE springs directly from the gradient function itself. Sometimes called the *vertex direction method*, it simply consists of finding the point ϕ^* that maximizes $D_{Q_c}(\phi)$, forming the one parameter family

$$(1-\gamma)Q_c + \gamma\Delta_{\phi^*}$$

and doing a one-dimensional algorithm in this family (usually Newton–Raphson) to find the maximum along this line. From our earlier theory, we know that this must increase the likelihood. It can be shown that, in fact, it provides a sufficient increase at each step that the algorithm must converge to the NPMLE [Lindsay (1983a)]. (We should more precisely say it has been proven that the likelihood will increase to its maximum value.)

Moreover, this is an algorithm that does not need to keep track of support points because the gradient, the new support point and the new likelihood vector can be constructed from the current likelihood vector. This is fortunate because this algorithm can add a new support point with each iteration.

Intuitive as this may seem, it is generally a very bad idea to use this algorithm except as a supplement to another speedier algorithm. As Lindsay (1983a) indicated, it is sublinear in convergence and in fact becomes slower and slower as we approach convergence.

For this reason, Böhning (1985) suggested a simple alternative that was considerably faster in terms of number of iterations. Called the *vertex exchange method*, it also requires only the solution of a one-dimensional optimization problem at each stage. Although it also has the disadvantage that we must keep track of the current latent distribution, in contrast with the vertex direction method, it can eliminate one support point at each step, so the number of support points stays bounded, with new points replacing old points.

6.3.4. *Intrasimplex direction method.* Neither of the preceding methods could be considered adequately speedy to do many calculations of the NPMLE, such as we might need to construct profile likelihoods or do simulation studies. The problem is that we need to somehow use the fact that unicomponent models are highly correlated when their parameter values are similar, and we cannot take advantage of this multicollinearity without using methods of higher dimension.

For this reason Lesperance and Kalbfleisch (1992) suggested an algorithm of a multivariate type, in which one found the set of optimal weights for a convex combination of the current likelihood vector and the unicomponent likelihood vectors corresponding to the current local maxima of the gradient function. As we have remarked earlier, the log likelihood will be strictly concave in these weight parameters, so a reasonable quasi-Newton procedure to solve for optimal weights, subject to nonnegativity constraints, could be expected to be fast and reliable. In addition, this algorithm does not require storage of the current latent distribution (the number of support points could grow explosively), but just the fitted likelihood vector.

6.3.5. *Monotonicity.* An important piece of practical advice to students starting work in this area is that no matter what method is used, one should be monitoring the likelihood function. If one is working with a method guaranteed to increase the likelihood, such as the EM, then this is a check on the program; if not, it is a preventative for oscillatory behavior. Since there is just a single unique maximum, there is no advantage to having the algorithm search the space more thoroughly. Böhning, Schlattman and Lindsay (1992) have some suggestions, and the corresponding C.A.MAN program provides various step length options.

6.3.6. *Using the dual problem.* Yet another approach to finding the NPMLE is to switch to solving the dual problem, which is a straightforward optimization problem with linear constraints defining the feasible region. Lesperance and Kalbfleisch (1992) use a canned program for this optimization problem (SIP = semiinfinite programming) and found that it was quite competitive with their intrasimplex direction method. It is the author's intuition that this will often be the best approach to the problem. Table 6.1 reproduces a comparison made by Lesperance and Kalbleish. The "Sup Grad" column indicates the final calculation of the convergence criterion and the column "Δl" indicates the difference between the global maximum log likelihood and the likelihood at the last iteration of the algorithm. These columns will be discussed further when we discuss stopping rules.

6.4. Ideal stopping rules. One could obviously take the tack that one should iterate on an algorithm until the accuracy of the result approaches the limit of machine accuracy. However, this is not practical if one wants to do simulation studies or apply bootstrap methods. We therefore attempt now to describe how to develop statistically meaningful stopping rules.

6.4.1. *The ideal rule.* We postulate that the *ideal stopping rule* for the iterations of an algorithm is to stop when we have a log likelihood that is sufficiently close to the final log likelihood. That is, we quit when we have found Q_{stop} satisfying

$$\ln(L(\hat{Q})) - \ln(L(Q_{\text{stop}})) \leq \text{tol}. \tag{6.1}$$

Here the criterion tol should be related to desired inferential goals. We here suggest what we think are reasonable values based on certain heuristics.

First, we note that setting such a goal will ensure that Q_{stop} will be a consistent estimator of Q, as was shown by Kiefer and Wolfowitz (1956).

Owen (1988) showed that one can construct valid nonparametric confidence sets for the smooth functionals of an unknown distribution function F by examining a form of nonparametric profile likelihood called the empirical

TABLE 6.1
A comparison of algorithms

Algorithm	**# Iterations**	**Sup Grad**	**Δl**	**APL Time**
VDM	2,177	2.64×10^{-3}	1.07×10^{-3}	12:28
VEM	143	3.18×10^{-2}	0.85×10^{-2}	1:04
ModVDM	60	4.96×10^{-4}	4.98×10^{-4}	5:49
SIP	11	1.60×10^{-8}	0	[0:14]
ISDM	11	1.44×10^{-8}	0	0:03

likelihood. Moreover, the empirical likelihood can be calibrated by the usual chi-squared distribution with degrees of freedom equal to the number of functionals under consideration.

It has not been shown that this theory carries over to our situation, but it seems likely to be true. Even if not, it gives us some guidance as to orders of magnitude that might be relevant. In our setting this would work as follows. Let $\theta = \theta(Q)$ be a functional of Q of interest, such as the mean or the distribution function evaluated at a particular point. Let $\hat{\theta} = \theta(\hat{Q})$. Let $\hat{Q}_\theta$ be the maximum likelihood estimator of Q among all Q that satisfy $\theta(Q) = \theta$. (We will discuss the solution of this optimization problem in the next chapter.) If Owen's result were to hold here, we could say, approximately, that

$$2[\ln(L(\hat{Q})) - \ln(L(\hat{Q}_\theta))] \approx (\hat{\theta} - \theta)[\operatorname{Var}\hat{\theta}]^{-1}(\hat{\theta} - \theta) \tag{6.2}$$

in the neighborhood of the maximum.

Now, suppose we set as a *target of accuracy* that our estimator of θ be within 0.1 standard error of $\hat{\theta}$. This is just 1/40 of the width of a standard confidence interval and so we believe it would be pointless to pursue numerical accuracy further than this, given the statistical inaccuracy. If we set

$$\text{tol} = 0.005, \tag{6.3}$$

stop at Q_{stop} and let θ_{stop} be the value of $\theta(Q_{\text{stop}})$, then

$$2[\ln(L(\hat{Q})) - \ln(L(\hat{Q}_{\theta_{\text{stop}}}))] \leq 2[\ln(L(\hat{Q})) - \ln(L(Q_{\text{stop}}))] \leq 0.01.$$

It follows, given our approximation (6.2), that θ_{stop} deviates from $\hat{\theta}$ by at most 0.1 standard units.

Thus we believe tol $= 0.005$ is a meaningful statistical goal and going beyond it pursues statistically meaningless accuracy.

6.4.2. *A gradient-based rule.* Now, we obviously cannot at any stage in an algorithm know exactly whether we have met the ideal stopping rule (6.1). However, we remind the reader that we can use the gradient to bound such a difference. That is, we know from Section 5.3.2 that

$$[\ln(L(\hat{Q})) - \ln(L(Q))] \leq \sup_{\phi\in\Omega} D_Q(\phi). \tag{6.4}$$

Thus if we can ensure that

$$\sup_{\phi\in\Omega} D_Q(\phi) \leq \text{tol},$$

we would have our targeted statistical accuracy.

One important question here regards whether the inequality (6.4) is close to an equality, at least in order of magnitude; otherwise we may be pushing our accuracy goals substantially beyond that actually needed. One piece of evidence for this is in Lesperance and Kalbfleisch (1992), in a table reproduced as Table 6.1 above. Our interest is in comparing columns 3 and 4, where we find the sup gradient and remaining likelihood to be gained at the final steps of each of the algorithms. We note that the sup gradient upper bound result shows that the last two algorithms have converged to a high degree of accuracy, so their likelihoods were set to be the true maximum. The numbers for the other estimators suggest that the order of magnitude of the remaining likelihood increase can be predicted fairly well from the sup gradient. In the worst case, the sup gradient was four times the remaining likelihood increase.

6.4.3. *Combining grid and gradient.* We now return to the supposition that we will evaluate the gradient at a finite grid of points. The next question we ask is: can we select a grid Ω_s and a modified tolerance level tol* in such a way that

$$\sup_{\phi \in \Omega_s} D_Q(\phi) \leq \text{tol}^* \quad \Longrightarrow \quad \sup_{\phi \in \Omega} D_Q(\phi) \leq \text{tol}? \tag{6.5}$$

Such a bound cannot be created unless the gradient has boundable variation, so that knowledge of its values on the grid points determines how high it can go between grid points.

This is not true in some cases, such as the empirical CDF problem, where one must include all the data points in the grid or else have zero likelihood for the data. In fact, as we will see, the larger the second derivatives of the unicomponent density are, the more refined must the grid be to attain desired statistical accuracy.

To analyze this question, we restrict attention to real-valued ϕ. Suppose that Q satisfies

$$\sup_{\phi \in \Omega_s} D_Q(\phi) \leq \text{tol}^*$$

and suppose that g_1 and g_2 are two adjoining grid points. Suppose that we can construct a bound of the type

$$\inf_{\phi \in [g_1, g_2]} D''_Q(\phi) \geq -c. \tag{6.6}$$

This ensures that the gradient cannot go upward and then curve downward too fast, and so bounds the maximum via

$$\sup_{\phi \in [g_1, g_2]} D_Q(\phi) \leq \text{tol}^* + \frac{c}{2}\left(\frac{g_2 - g_1}{2}\right)^2. \tag{6.7}$$

(The reader is invited to check this out: The bound arises by setting the gradient equal to tol* at the two endpoints and then making the steepest possible quadratic in between.)

Thus, it is clear that if we can find c to satisfy (6.6), then by application of (6.7), together with a fine enough grid, one can attain the goal (6.5).

6.4.4. *Bounding the second order score.* One of the problems with establishing a bound of the type (6.6) is the presence of the arbitrary distribution Q as an argument. Thus we will modify the above strategy somewhat to make the problem easier to solve. We remind the reader that if $L_i(\phi) = f(x_i;\phi)$, then the dispersion score

$$v_2(\phi, x_i) = \frac{f''(x_i;\phi)}{f(x_i;\phi)} = \frac{L_i''(\phi)}{L_i(\phi)}$$

played an important role in evaluating overdispersion.

The following lemma suggests that this score is an important quantity in determining the properties of the gradient as well.

LEMMA 26. *If for all i and for all $\phi \in [g_1, g_2]$,*

$$\frac{L_i''(\phi)}{L_i(\phi)} \geq -k, \tag{6.8}$$

then

$$D_Q''(\phi) \geq -k \left[n + \sup_{\phi \in [g_1, g_2]} D_Q(\phi) \right]. \tag{6.9}$$

PROOF. Straightforward algebra, using the fact that

$$n + D_Q(\phi) = \sum n_i \frac{L_i(\phi)}{L_i(Q)}.$$ □

Before proceeding, we note that it is easy to establish a bound such as (6.8) in the exponential family. For example, if ϕ is the natural parameter, we have

$$\frac{L_i''(\phi)}{L_i(\phi)} = v_2(\phi, x_i) = [x_i - \mu(\phi)]^2 - \sigma^2(\phi),$$

which is clearly bounded below by $-\sigma^2(\phi)$ on the chosen interval. If we are using instead the mean value parameterization, then

$$\frac{L_i''(\phi)}{L_i(\phi)} = \frac{(x_i - \mu(\phi))^2 - \sigma^2(\phi)}{\sigma^4(\phi)} \geq -\frac{1}{\sigma^2(\phi)}$$

is a simple bound on an interval in which the unicomponent variance of X does not go to zero.

6.4.5. *A conservative method.* Now if we insert (6.9) into (6.5), we get

$$\sup_{\phi \in [g_1, g_2]} D_Q(\phi) \leq \text{tol}^* + k \left(\frac{(g_2 - g_1)^2}{8} \right) \left[n + \sup_{\phi \in [g_1, g_2]} D_Q(\phi) \right].$$

We let

$$C = k \left(\frac{(g_2 - g_1)^2}{4} \right)$$

be the *critical factor* for the grid. For example, if we use the mean value parameter of the exponential family, we get

$$C = \frac{w^2}{\sigma^{*2}},$$

where w is the half-width of the grid separation and σ^{*2} is a lower bound for $\sigma^2(\phi)$ on that interval. Algebraic manipulation then gives the upper bound

$$\sup_{\phi \in [g_1, g_2]} D_Q(\phi) \leq \frac{\text{tol}^* + Cn/2}{1 + C/2}.$$

Clearly for C sufficiently small, one can make this bound as close to tol* as one likes. Moreover, for C small, the bound is approximately $\text{tol}^* + Cn/2$.

We note that the choice of an acceptable critical factor C for the grid will therefore depend very much on the sample size n. This derives from the fact that as the sample size increases, the greater the precision of our statistical knowledge and so the greater the need for numerical accuracy. Thus we run into the unfortunate side effect that *the larger the sample size, the more difficult the numerical problem.*

Returning to the exponential family in the mean value parameterization, we see that the critical factor for the grid is the grid separation expressed in standard deviation units. Thus, for example, if we were to separate the grid points by 0.02 standard deviations, we would get an upper bound of the form

$$\sup_{\phi \in \Omega} D_Q(\phi) \leq \frac{\text{tol}^* + (0.00005)n}{1 + (0.00005)}.$$

In this case, the standardized grid widths must shrink at the rate $n^{-1/2}$ to maintain the desired accuracy. For example, if our target tolerance is 0.005, as suggested earlier, and we set tol* for the grid at 0.0025, then we must set

$$\frac{w}{\sigma^*} = \sqrt{0.005}n^{-1/2}.$$

Thus for $n = 100$, we need a standardized half-grid-width of about 0.007. For $n = 10{,}000$, the grid half-width shrinks to 0.0007 standardized units.

6.4.6. *Remarks.* We remark that this analysis indicates that if we were to change to an exponential family parameterization in which the variance is constant, then we could use an equally spaced grid without losing accuracy in any region, but otherwise there will be a potential loss of information in the use of an equally spaced grid.

These design considerations have been based on the idea of conservatism and least favorable situations. Empirical evidence is not available on whether these recommendations are overly conservative.

We also note that if one is doing a gradient search based on a grid of starting points, that these considerations suggest that the grid should be evenly spaced on the standardized scale, and that the spacing should shrink with sample size.

CHAPTER 7

Extending the Method

In this chapter we present some simple extensions of the NPMLE theorem that solve problems that are similar, but not identical, in structure. We consider three situations: first, a class of problems in which the unknown latent distribution appears in the likelihood in a ratio form; second, the question of maximizing a mixture likelihood with linear constraints on the latent distribution; third, the problem of estimating the latent distribution with a continuous density function.

7.1. Problems with ratio structure. We start with a simple example that illustrates a problem in which the unknown latent distribution shows up in the likelihood in a ratio form.

7.1.1. *Example: Size bias.* Suppose that $X_1, \ldots, X_n$ are positive-valued random variables, but they arose from a population that was sampled not randomly, but with probabilities that are proportional to some positive function $w(x)$ of the variable of interest. That is, suppose the underlying distribution of the variable X is G, with density g, but the sampling is from the density proportional to $w(x)g(x)$.

A classic example of this type would be if we were to sample vacationers in a hotel lobby and ask how long they were staying in the hotel. The vacationers who have longer stays are more likely to be included in the sample.

The nonparametric MLE problem is then to find the underlying distribution G given knowledge of the sampling weights $w(x)$. We can write the likelihood, for discrete G, as

$$L(G) = \prod \frac{w(x_i)G(\{x_i\})}{\int w(x)\, dG(x)} = \prod \frac{\int w(x_i)\mathscr{I}[\phi = x_i]\, dG(\phi)}{\int w(x)\, dG(x)}.$$

The question is, how do we maximize such a likelihood, which now has the latent distribution in both numerator and denominator?

7.1.2. *NPMLE with ratio structure.* We first generalize the problem. We desire to solve a problem of the form

$$\sup_G \prod_{i=1}^{n} \left[\frac{\int h_i(\phi)\, dG(\phi)}{\int H(\phi)\, dG(\phi)} \right]. \tag{7.1}$$

We require that $H(\phi)$ be positive-valued. The first step is to reparameterize the problem by forming a reweighted version of G:

$$dQ(\phi) = \frac{H(\phi)\, dG(\phi)}{\int H(\phi)\, dG(\phi)}. \tag{7.2}$$

If this is done, check that we can rewrite the original problem as

$$\sup_Q \prod \left[\int \frac{h_i(\phi)}{H(\phi)}\, dQ(\phi) \right]. \tag{7.3}$$

Now we note that this is exactly of the mixture NPMLE form, where we use the likelihood kernels $L_i(\phi) = h_i(\phi)/H(\phi)$. Thus we can solve this problem to find a maximizing $\hat{Q}$. To find the NPMLE for the original problem, we must undo the transformation (7.2), and we obtain the following result:

PROPOSITION 27. *If $\hat{Q}$ solves the modified problem* (7.3), *then*

$$d\hat{G}(\phi) = \frac{[1/(H(\phi))]\, d\hat{Q}(\phi)}{\int [1/(H(\phi))]\, d\hat{Q}(\phi)}.$$

solves the original problem (7.1).

7.1.3. *Example: Size bias.* In the size bias problem, we have $H(\phi) = w(\phi)$ and

$$h_i(\phi) = w(\phi)\mathscr{I}[\phi = x_i].$$

It follows that the likelihood kernels for the transformed problem are

$$L_i(\phi) = \frac{h_i(\phi)}{H(\phi)} = \mathscr{I}[\phi = x_i].$$

Therefore the transformed problem is exactly that for which the empirical CDF, mass n^{-1} at each x_i, is the solution $\hat{Q}$. Following through the next step, we obtain the standard solution to the weighted sampling problem, that $\hat{G}$ has the form

$$\sum_i \left[\frac{w(x_i)^{-1}}{n \sum_j w(x_j)^{-1}} \right] \Delta_{x_i}.$$

7.1.4. *Example: Weibull competing risks.* We introduce, with a motivating Weibull example, another class of problems in which the maximum likelihood solution requires solving a ratio problem. Suppose we have a piece of machinery with an unknown number of independent parts that are each subject to failure with a Weibull lifetime distribution. We observe the first time to failure. Thus we are in a setting where there are an unknown number of competing risks, assumed to be independent. We also do not observe the cause of failure.

We let $T_1, \ldots, T_v$ be the latent failure times of the competing sources of risk and so

$$X = \min\{T_1, \ldots, T_v\}$$

is the observed random variable. Since the latent times are independent, the *cumulative hazard function* of the observed variable, say $H(t)$, is the sum of the individual cumulative hazards for the latent variables, in the Weibull case, $H_r(t) = \lambda_r t^{\alpha_r}$. That is,

$$H(t) = \sum_{r=1}^{v} H_r(t) = \sum_r \lambda_r t^{\alpha_r}.$$

We can put this into a mixture format by setting

$$\Lambda = \sum \lambda_r$$

and defining the discrete distribution G to have mass λ_r/Λ at α_r. If this is done, we can write the cumulative hazard function in the form

$$H_X(t) = \Lambda \cdot \int t^\alpha \, dG(\alpha).$$

Thus we can see that if we have a problem with an unknown number of competing Weibulls, we have a *mixed hazard function*. We note also that if there are multiple latent failure variables with the same value of α, we cannot identify the separate λ_i from this hazard function because they show up in the hazard only through their total. We also note that the hazard intensity rate has a mixed form:

$$h_X(t) = \frac{d}{dt} H_X(t) = \Lambda \cdot \int \alpha t^{\alpha-1} \, dG(\alpha).$$

We would like to maximize the likelihood of a sample of observations as a function of the unknown (Λ, G).

7.1.5. *Mixed hazards NPMLE.* There is an obvious generalization of the Weibull formulation in which we have a family of distributions whose cumulative hazard can be expressed as a mixture of kernel cumulative hazards,

$$H_X(t) = \Lambda \cdot \int K(t;\alpha) \, dG(\alpha) = \Lambda \cdot K(t;G),$$

and whose hazard rate therefore has the form

$$h_X(t) = \Lambda \cdot \int k(t;\alpha) \, dG(\alpha) = \Lambda \cdot k(t;G),$$

where k is the derivative of K with respect to t. Such a formulation can arise from a competing risk framework, as in the Weibull example.

However, this formal structure can also arise if we partition the time axis into regions A_r of unknown but constant hazard, writing kernel hazards of the form $k(t;r) = \mathscr{I}[t \in A_r]$, and obtaining the mixed hazard model

$$h_X(t) = \sum \lambda_r \mathscr{I}[t \in A_r].$$

In this case the "latent variable" α is discrete, corresponding to the index r of the interval, and the setting is parallel to the mixture problem with known component distributions.

We now consider the likelihood for the problem, referring the reader to the paper by Hsi, Lindsay and Lynch (1992) for details on how to incorporate censoring. We recover the density for X from the hazard specification as

$$f_X(t) = h_X(t) \exp(-H_X(t)).$$

It follows that the likelihood for a sample has the form

$$L(\Lambda, G) = \exp(-\Lambda \sum K(x_i; G))\, \Lambda^n \prod k(x_i; G).$$

We next fix G and maximize the likelihood over Λ to find

$$\hat{\Lambda}_G = n\left(\sum K(x_i; G)\right)^{-1} = \bar{K}_G^{-1}.$$

It follows that the profile likelihood for the unknown distribution G has the form

$$L(\hat{\Lambda}_G, G) = e^{-n} \prod \left[\frac{k(x_i; G)}{\bar{K}_G}\right].$$

However, this problem is exactly of the ratio type we have described before and so we can solve it from the NPMLE theorem by transformation.

We note that Hsi, Lindsay and Lynch (1992) obtained this solution through a more difficult route, and so this presentation has its value in showing the simple structure that lies behind the problem. See the cited paper for further details on the applications of this model.

7.2. NPMLE with constraints on Q. There are a number of circumstances in which we might wish to maximize the nonparametric mixture likelihood under constraints on the unknown latent distribution Q. We might, for instance, need for identifiability reasons to constrain it to have mean 0 and variance 1. Another situation in which we might want to employ constraints is as follows.

7.2.1. *Profile likelihood.* Suppose we wish to form a profile likelihood based on some function of Q that is of interest. To be specific, let us say the mean value of the latent distribution is of interest:

$$\tau(Q) = \int \phi \, dQ(\phi).$$

To form a profile likelihood $\mathbb{L}$ in such a nonparametric setting, we calculate for each fixed value of τ_0, the solution to a constrained maximum likelihood problem,

$$\mathbb{L}(\tau_0) = \sup\{L(Q)\colon Q \ni \tau(Q) = \tau_0\}. \tag{7.4}$$

The results of Owen (1988) suggest that such a *nonparametric profile likelihood* might well give us a method of performing tests and constructing confidence intervals for sufficiently smooth functions $\tau(Q)$ of the latent distribution.

Although the asymptotic theory is not yet available, we believe that profile likelihood intervals will provide a value tool for understanding which features of the latent distribution are trustworthy and which are poorly determined by the data. If profile likelihood intervals are carried out with a good optimization routine, they may be substantially more time efficient than bootstrapping and also provide a more natural way to construct confidence sets in more than one dimension.

Thus we have a modified optimization problem, and our goal here is to provide a theory for its solution.

7.2.2. *Linear constraints.* In this chapter we will consider only linear constraint problems. One type of such constraint is the *linear equality constraint*, by which we mean there are a set of functions $h_1(\phi), \ldots, h_a(\phi)$ and a set of constants $h_1^o, \ldots, h_a^o$ such that we wish to maximize the likelihood subject to the following restrictions on Q:

$$\begin{aligned} \int h_1(\phi)\, dQ(\phi) &= h_1^o \\ \vdots \qquad &\quad \vdots \\ \int h_a(\phi)\, dQ(\phi) &= h_a^o. \end{aligned} \tag{7.5}$$

We can add yet more flexibility in fitting profile likelihoods by allowing an additional set of *linear inequality constraints*:

$$\begin{aligned} \int k_1(\phi)\, dQ(\phi) &\le k_1^o \\ \vdots \qquad &\quad \vdots \\ \int k_b(\phi)\, dQ(\phi) &\le k_b^o. \end{aligned} \tag{7.6}$$

7.2.3. *Examples with linear constraints.* As examples of the equality constraint, the moments of the distribution of Q are the most obvious parameters of interest. However, we can also construct a profile likelihood for the distribution function of Q at a particular fixed value, say $Q\,(\phi_0)$, by using an indicator function

$$h_1(\phi) = \mathscr{I}[\phi \le \phi_0]$$

to construct the linear equality constraint.

Here is another example that is less obvious: Suppose we are interested in constructing a confidence interval for a posterior empirical Bayes function, such as

$$E[g(\Phi) \mid X = x_0] = \frac{\int g(\phi) f(x_0;\phi)\, dQ(\phi)}{\int f(x_0;\phi)\, dQ(\phi)}.$$

(Note that x_0 is a constant here.) Although this posterior mean is not a linear function of Q, when its value is fixed at some constant c, the maximization takes place over a set of Q defined by the following linear equality constraint:

$$\int [g(\phi) f(x_0;\phi) - cf(x_0;\phi)]\, dQ(\phi) = 0.$$

If we wish to do inference on the quantiles of Q, we can turn to the linear inequality constraints version of the problem. For example, if we fix the median of Q to be a specified value, say c, then we can maximize the likelihood subject to the two simultaneous linear inequalities:

$$\int \mathscr{I}[\phi \le c]\, dQ(\phi) \ge 0.5,$$
$$\int \mathscr{I}[\phi \ge c]\, dQ(\phi) \ge 0.5.$$

In this fashion, one could construct the profile likelihood of the median of Q.

7.2.4. *The constrained NPMLE.* We now consider the properties of the NPMLE if there are equality and inequality constraints in the form (7.5) and (7.6). Because the constraints are linear, we can extend our previous geometric analysis to allow for their consideration. We construct an extended likelihood vector in $(D + a + b)$-dimensional Euclidean space,

$$\mathbf{L}^*(\phi) := \begin{pmatrix} \mathbf{L}(\phi) \\ \mathbf{h}(\phi) \\ \mathbf{k}(\phi) \end{pmatrix},$$

and we let its convex hull be $\mathscr{M}^*$. The elements of $\mathscr{M}^*$ are of the form

$$\begin{pmatrix} \mathbf{L}(Q) \\ \mathbf{h}(Q) \\ \mathbf{k}(Q) \end{pmatrix}.$$

Let

$$\mathbf{x} = \begin{pmatrix} \mathbf{p} \\ \mathbf{q} \\ \mathbf{r} \end{pmatrix}$$

be an arbitrary vector of dimension $D + a + b$, and define the log likelihood objective function on this space by

$$l(\mathbf{x}) = \sum_{i=1}^{D} n_i \ln(p_i),$$

noting that the last $a+b$ coordinates are given zero weight, so that the objective function is not strictly concave any longer.

Next, we define the suitable set over which the optimization will take place. The set of latent distributions that give extended likelihood vectors that meet our constraints lie in the set

$$\mathcal{M}^*_{\text{cons}} = \mathcal{M}^* \cap \{\mathbf{x}: \mathbf{q} = \mathbf{h}^o\} \cap \{\mathbf{x}: \mathbf{r} \leq \mathbf{k}^o\}.$$

The observation to be made here is that $\mathcal{M}^*_{\text{cons}}$ is a convex set, and so we can establish some results directly about this optimization problem.

PROPOSITION 28. *If the likelihood vector curve is closed and bounded and the mixture set contains points of positive likelihood, then:*

1. *If there are only linear equality constraints, then there exists a unique maximizing vector* $\hat{\mathbf{L}}$.
2. *If there are both equality and inequality constraints, then the maximum likelihood vector may not be unique, but there does exist a convex set of maximum likelihood solutions.*
3. *In either case, all solutions can be represented as a mixture with* $D + a + b$ *or fewer components.*

PROOF. For part 1, we note that the equality constraints form $\mathcal{M}^*_{\text{cons}}$ by taking a slice through $\mathcal{M}^*$. The resulting cross section set is still of dimension D and the objective function is still strictly concave on this set. The strict concavity gives the uniqueness. Concavity implies the presence of the solution on the boundary of $\mathcal{M}^*_{\text{cons}}$ and hence the boundary of $\mathcal{M}^*$.

For part 2, we lose strict concavity of the objective function on $\mathcal{M}^*_{\text{cons}}$ when we have inequality constraints, because the coordinates of $\mathbf{x}$ corresponding to the inequality constraints can now be varied without affecting the objective function. Under nonstrict concavity, we can only make the weaker statement.

Part 3 follows from the fact that the solutions must lie in the boundary of the set $\mathcal{M}^*$, so we can apply Carathéodory's theorem with the dimension reduced by one. □

7.2.5. *A simple algorithm.* Problems of this type can be solved by the technique of Lagrange multipliers. We will describe here the case where all constraints are of the linear equality type. Our problem is then to jointly maximize over Q and the Lagrange multipliers $\lambda_1, \ldots, \lambda_a$ the objective function

$$\sum n_i \ln(L_i(Q)) + \lambda_1[h_1(Q) - h_1^o] + \cdots + \lambda_a[h_a(Q) - h_a^o].$$

Our approach will be to treat the Lagrange multipliers $\boldsymbol{\lambda}$ as fixed, and to maximize just over Q at first. We can readily find the gradient function for this new criterion to be

$$D_{Q,\lambda}(\phi) = D_Q(\phi) + \sum \lambda_r[h_r(\phi) - h_r(Q)]. \tag{7.7}$$

With this gradient function, we are ready to solve the problem. We can either use a straightforward gradient algorithm from this point, or we can use the EM algorithm with constraints, and just use this gradient function to check the convergence. Our output will be a mixture solution $\hat{Q}_{\boldsymbol{\lambda}}(\phi)$. Note that the Lagrange gradient (7.7) does *not* depend on the initial conditions specified, but only the multipliers $\boldsymbol{\lambda}$ themselves, and so the same is true of $\hat{Q}_{\boldsymbol{\lambda}}(\phi)$.

Our next observation is that even though this $\hat{Q}_{\boldsymbol{\lambda}}(\phi)$ need not satisfy our initial constraints, we can easily solve for the set of linear constraints that it actually does satisfy:

$$\begin{array}{ccc} \int h_1(\phi)\, dQ(\phi) = & \int h_1(\phi)\, d\hat{Q}_{\boldsymbol{\lambda}}(\phi) := & h_1^*(\boldsymbol{\lambda}) \\ \vdots & \vdots & \vdots \\ \int h_a(\phi)\, dQ(\phi) = & \int h_a(\phi)\ \hat{Q}_{\boldsymbol{\lambda}}(\phi) := & h_a^*(\boldsymbol{\lambda}). \end{array}$$

Checking backward, we find that if we had started with $h_1^*(\boldsymbol{\lambda}), \ldots, h_a^*(\boldsymbol{\lambda})$ as our initial constraints $h_1^o, \ldots, h_a^o$, then $\hat{Q}_{\boldsymbol{\lambda}}(\phi)$ and $\lambda_1, \ldots, \lambda_a$ would have been the solutions to the Lagrange multiplier problem.

That is, we did not obtain the value $\mathbb{L}(h_1^o, \ldots, h_a^o,)$ of the profile likelihood with our original $h_1^o, \ldots, h_a^o$, but we do end up with another value of the profile likelihood, namely,

$$\mathbb{L}(h_1^*, \ldots, h_a^*).$$

It follows that reconstruction of the profile likelihood can be carried out by interpolation over a selected region of Lagrange multiplier values, chosen to give h^* values in the right region. This operation is particularly simple if there is a single constraint function $h(\phi)$ which is nonnegative, because then $h^*(\lambda)$ is monotonically decreasing in λ. [*Exercise.*]

7.3. Smooth estimates of *Q*. This author is not, on the whole, in favor of the idea of using continuous densities to estimate the latent distribution. Although the argument that nature is continuous has some compelling features, in most models the level of information about the latent distribution is simply too small to consider any discrimination about the form of this distribution. In essence, as indicated in Chapter 1, one can obtain reasonable estimates or intervals only for the smoothest of functionals of the latent distribution, and the goal of actually discerning the true density is typically impossible for all practical purposes. However, there are certainly some examples, especially in image analysis, where the prior information that the picture is relatively smooth is important in performing a useful analysis.

Therefore, some further references in this regard will be given for the sake of readers with a particular statistical interest in a smooth estimator. In addition, at the end there will be a suggestion of my own about how to directly use the nonparametric likelihood approach and still obtain a smooth estimator by maximum likelihood.

7.3.1. *Roughening by smoothing.* One approach that has been taken is to take a very smooth initial estimator, such as normal latent distribution, and apply the EM algorithm to it for a few steps. The EM algorithm formula generalizes very simply to the gradient updating of density functions by the formula

$$q_{\text{em}}(\phi) = q_c(\phi)[1 + n^{-1} D_Q(\phi)].$$

If we start with a normal density, the first step takes us to a mixture of n normal densities, each with smaller variance. Laird and Louis (1991) call this "smoothing by roughening." A similar approach is recommended by Vardi and Lee (1993).

7.3.2. *Deconvolution.* Another approach is to extend the idea of kernel-based density estimation into the domain of the latent distribution. There have been a number of papers in this regard, of which we might mention Fan (1991). The method usually relies on the convolution type mixture and is found in the literature under the keyword deconvolution. The most important lesson from this literature is that the best possible rates of convergence are extremely poor, and therefore density estimation is practically impossible.

7.3.3. *Series expansion.* Another set of workers have developed analogues of series expansions to use for fitting the latent distribution smoothly. Gallant and Nychka (1987) called this a semi-nonparametric approach. A similar approach is carried out by Walter and Hamedani (1991) in the context of empirical Bayes estimation.

7.3.4. *A likelihood method.* We can easily extend the nonparametric maximum likelihood idea to construct estimates with smooth densities possessing any prespecified degree of smoothness. In particular, we can choose the estimated density to have a likelihood nearly that of the global nonparametric maximum likelihood estimator.

We start with a family of densities $g(\phi; \theta, \tau)$ on the parameter space Ω. The parameter θ is assumed to determine the central location of this density and τ is a dispersion parameter, with the distribution concentrating about θ as $\tau \to 0$. Of course, the normal density is such a family, but in many situations the natural conjugate density family might be more suitable because it would avert numerical integration problems.

If we have a known component model, there might be a natural way to construct a kernel family that gives some target smoothness to the probabilities over physically neighboring components. For example, in positron emission tomography, one could construct a discrete distribution over the sites ϕ that are neighbors to site θ, with the dispersion parameter τ reflecting the amount of mass spread to the neighbors.

Ideally this construction is done so that we can explicitly calculate the marginal distribution of X when g is the latent distribution. That is, we desire

$$f^*(x; \theta_j, \tau) = \int f(x; \phi) g(\phi; \theta_j, \tau)\, d\phi$$

to be readily calculable. Note that under our specifications, as $\tau \to 0$ this density should go to the unicomponent density $f(x;\theta_j)$, and we so assume.

Suppose we replace our basic discrete class of latent distributions $\sum \pi_j \Delta_{\phi_j}$ with arbitrary convex combinations of latent distributions of the form

$$\sum \pi_j g(\phi;\theta_j,\tau).$$

The resulting class of mixture densities for X can now be expressed as convex combinations of new family of basic densities, namely,

$$X \sim \sum \pi_j f^*(x;\theta_j,\tau) = \int f^*(x;\theta,\tau)\, dH(\theta).$$

Now we can describe a strategy. For each fixed τ we can calculate the NPMLE for this new family of mixtures. If the answer is $\hat{H}_\tau$, then we have a resulting smooth density estimator for the original problem, namely,

$$q_\tau(\phi) = \int g(\phi;\theta,\tau)\, d\hat{H}_\tau(\phi),$$

with corresponding distribution $\hat{Q}_\tau$. The selection of τ can be based on likelihood considerations. The NPMLE for the problem necessarily has higher likelihood, but if we target a fixed difference

$$\ln L(\hat{Q}) - \ln L(Q_\tau) = \delta,$$

then consistency will follow from the Kiefer–Wolfowitz (1956) result.

Moreover, if we set δ sufficiently small, say 0.005, the arguments of Chapter 6 imply that the resulting estimator will differ very little from the NPMLE in inference for nonparametric functionals. However, this will come at the cost of losing much of the smoothness of the estimator.

CHAPTER 8

The Semiparametric MLE

Although there are certainly examples where interest in the mixture NPMLE method has focused on the estimated latent distribution itself or functionals of it, the largest class of investigations and applications has occurred in the arena of semiparametric estimation, in which the latent distribution is included in the model to allow extra heterogeneity, but the focus is on a set of auxiliary parameters, generally of the regression type.

The earliest extensive applied investigation of semiparametric mixture maximum likelihood was by Heckman and Singer (1984), who were investigating the effects of latent distribution misspecification and were comparing, therefore, the nonparametric and parametric approaches to modeling the latent distributions. Among the many further investigations of this type we might count Follman and Lambert (1989), Brännäs and Rosenqvist (1994), Butler and Louis (1992), and Davies (1993). See Lindsay and Lesperance (1995) for a survey of the results in this area.

Since there exists a substantial literature on the implementation of this methodology, we will focus herein instead on special simplifying structures that can exist when we have a likelihood with both latent and auxiliary parameters. We will consider three different semiparametric models. One is an exponential family random effects model, of which the Rasch model is an illustration. The second is a measurement error problem, with the additional complication of case-control sampling. The final problem is an outlier distribution model for contingency tables that leads to a new method for assessing the fit of a parametric model.

We start with a general optimization result that is important because it leads to a simplification and clarification of what would otherwise be numerically more difficult problems. It will be used in the examples.

8.1. An equivalence theorem. In this section, we will use the notation β and G to designate two arbitrary "parameters" upon which our model depends, but in our applications, β will be real-valued parameters of interest and G will be the latent distribution. Our interest is in whether the β parameters we would obtain from maximizing a conditional likelihood will be the same as we would obtain from maximizing a simpler joint likelihood.

The setting: We wish to maximize a target likelihood L_c with ratio structure:

$$L_c(\beta, G) = \frac{L_J(\beta, G)}{L_m(\beta, G)}.$$

We let $(\hat{\beta}_c, \hat{G}_c)$ be maximizers of L_c and we wish to compare them with $(\hat{\beta}_J, \hat{G}_J)$, which will come from maximizing the simpler likelihood L_J. In the settings in which this is applied, the likelihood we wish to maximize, L_c, is a *conditional* likelihood. Then the numerator likelihood L_J is the *joint* likelihood, which is of the desired product form, while the denominator likelihood is the *marginal* likelihood L_m of the conditioning statistics.

Our first key assumption is that the denominator likelihood has multinomial structure; that is, there are constants $n_1, \ldots, n_K$ and nonnegative functions $p_j(\beta, G)$ satisfying

$$\sum p_j(\beta, G) = 1$$

such that the likelihood can be written

$$L_m(\beta, G) = \prod p_j(\beta, G)^{n_j}.$$

Let $n = \sum n_j$. This structure will clearly arise if we are conditioning on a set of discrete statistics.

The next assumption specifies a form of *nonidentifiability* of the parameter G in the conditional likelihood. That is, we assume that for every pair (β, G) in the parameter space there exists G^* such that

$$L_c(\beta, G) = L_c(\beta, G^*) \tag{8.1}$$

and

$$p_j(\beta, G^*) = \frac{n_j}{n} \qquad \forall j. \tag{8.2}$$

That is, no matter which β we start with, we can find G^* that shifts the multinomial probabilities under the model to equal the observed sample proportions without altering the target likelihood. We note that this is where the nonparametric nature of G is important, so that there are "sufficient degrees of freedom" to solve these equations.

With these assumptions, we are done, because as far as the parameter of interest β is concerned, we can equally well maximize either L_J or L_c; the solutions are the same.

PROPOSITION 29. *Under the preceding assumptions, if* $(\hat{\beta}_c, \hat{G}_c)$ *maximizes the target likelihood* L_c, *then* $(\hat{\beta}_c, \hat{G}_c^*)$ *maximizes* L_J. *Conversely, if* $(\hat{\beta}_J, \hat{G}_J)$ *maximizes* L_J, *then* $(\hat{\beta}_J, \hat{G}_J)$ *satisfies* (8.2) *and is in an equivalence class that maximizes* L_c.

PROOF. We write the likelihoods in the product form

$$L_J(\beta, G) = L_c(\beta, G) L_m(\beta, G).$$

We note that by the assumption (8.1), if $(\hat{\beta}_c, \hat{G}_c)$ maximizes L_c, then so must $(\hat{\beta}_c, \hat{G}_c^*)$. Further, $(\hat{\beta}_c, \hat{G}_c^*)$ also maximizes L_m because of its multinomial form and (8.2). The sample proportions give, in fact, the maximum possible marginal likelihood over all possible models. As to the converse, if (8.2) was not satisfied at the maximum, we could increase the likelihood component L_m without changing that of L_c by choosing G^* to make it so, a contradiction. If $(\hat{\beta}_J, \hat{G}_J)$ did not maximize the other component L_c, then we could increase L_c without decreasing L_m by the same device. □

8.2. Exponential response models. We now consider a simple class of exponential response models. A sample is taken from a population, with the measurement being a vector $x = (x_1, \ldots, x_r)'$, with x_i, being the "response" to the ith out of r "items." There will be item parameters $\theta = (\theta_1, \ldots, \theta_r)$ that determine the distribution of responses and a parameter ϕ_i that reflects the latent propensities of the sampled unit. The density for the ith sampling unit, conditional upon latent variable $\Phi = \phi_i$, will have the exponential form

$$f(x_i; \theta, \phi_i) = \exp[\theta' x + \phi_i s(x) - k(\theta, \phi_i)],$$

where $s(x)$ is the sufficient statistic for the variable ϕ_i.

For the ensuing discussion, it will be essential that $s(x)$ has a finite discrete distribution, say with sample space $\{0, \ldots, K\}$. [A subject for investigation would be its approximate validity when $s(x)$ is continuously distributed.]

Two other features of the exponential form that are important in the theory are:

1. The statistic $s(x)$ is complete and sufficient for ϕ when θ is fixed;
2. The marginal distribution of $s(x)$ is an exponential family when θ is fixed.

8.2.1. *Example: Rasch model.* A model of the type we are interested in that has received a great deal of attention is the *Rasch model*, in which the responses X_{ij} are binary variables. We start with a logistic model for the ith subject's response to the jth item, conditional on the subject's latent variable $\Phi = \phi_i$:

$$\Pr[X_{ij} = 1 \mid \Phi_i = \phi_i] = \frac{\exp(\theta_j + \phi_i)}{1 + \exp(\theta_j + \phi_i)}.$$

We further assume that conditional on the latent variable, all the responses of an individual are independent, so that conditional on the latent variable the density has the exponential response form

$$\Pr[X_i = x_i \mid \Phi = \phi_i] = \exp\left(\sum_j \theta_j x_{ij} + \phi_i x_{i\cdot} - \kappa(\theta, \phi_i)\right),$$

where the sufficient statistic for ϕ_i is the response total $x_{i\cdot}$ of the ith subject. The model is overparameterized, so to make things identifiable a constraint must be used. We will here specify that the last item parameter, θ_K, is zero.

8.2.2. *Type I conditional models.* A reason for particular interest in the exponential response model is that if the focal parameters are the θ's, then there exists a natural competitor to using semiparametric mixture likelihood methods. Because of the structure of the model, one can form a *conditional likelihood* for each subject that depends strictly on the parameters of interest, namely,

$$L_{i,\text{cond}}(\theta) = \Pr[X_i = x_i \mid s(X_i) = s_i].$$

We can now estimate the focal parameters without regard to the structure of the latent variables, whether they are treated as nuisance parameters or as a sample from an unknown distribution. The resulting conditional maximum likelihood estimators (or maximum conditional likelihood estimators) are very generally consistent and highly efficient relative to maximum likelihood [e.g., Liang (1984)]. Important early work on this method was done by Anderson (1973).

8.2.3. *The two-item example.* A simple example that illustrates our situation is the *paired Bernoulli* problem, which can also be described as the Rasch model with two items. We have from each unit (subject or paired subjects) two binary variates (X_{i1}, X_{i2}). We might think of the first as corresponding to a response under treatment and the second as the response under control. We apply the Rasch model, where we have one item parameter $\theta = \theta_1$ (since $\theta_2 = 0$), which corresponds to the log odds ratio for the success parameters for the pair, and so represents a common treatment effect. There are only four possible responses for each pair, namely, $(0,0)$, $(1,0)$, $(1,0)$ and $(1,1)$. We let $N(a,b)$ denote the number of pairs of responses with pattern (a,b).We can therefore easily write the conditional densities for the data given the sufficient statistic $X_{i1} + X_{i2} = S_i$:

$$\begin{aligned}
\Pr[X = (0,0) \mid S = 0] &= 1,\\
\Pr[X = (1,0) \mid S = 1] &= \frac{\exp(\theta)}{1+\exp(\theta)},\\
\Pr[X = (0,1) \mid S = 1] &= \frac{1}{1+\exp(\theta)},\\
\Pr[X = (1,1) \mid S = 2] &= 1.
\end{aligned}$$

The conditional likelihood for the problem is therefore

$$L_{\text{cond}}(\theta) = \left[\frac{\exp(\theta)}{1+\exp(\theta)}\right]^{N(1,0)} \left[\frac{1}{1+\exp(\theta)}\right]^{N(0,1)}$$

This is easily maximized for θ and provides a method of consistent estimation even though the fixed effects MLE is inconsistent. However, it offers a curiosity in that it only uses the information from the *discordant pairs*; that is, those in which the binary variates were not equal. The estimator does not use $N(1,1)$ or $N(0,0)$, the numbers of concordant pairs.

8.2.4. *Efficiency theorem.* My early interest in the mixture problem arose around the statistical question of the efficiency of the conditional approach to eliminating nuisance parameters. Addressing the question of efficiency in the case of the number of nuisance parameters going to infinity is quite difficult. My approach was to consider the efficiency of the conditional approach within the semiparametric mixture model, where the nuisance parameters ϕ_i were assumed to come from a latent distribution Q. The problem is still difficult, but there is sufficiently more structure so that one can, in a natural way, extend the efficiency ideas from the parametric model to the nonparametric, and answer this question.

The answer [Lindsay (1983c)] is that the conditional method is generally fully efficient within the mixture setting. This last paper was written when the appropriate semiparametric theory was nascent; for an up-to-date description of the necessary optimality theory, we refer the reader to the book by Bickel, Klaassen, Ritov and Wellner (1993).

We note that there are some technical issues concerning the boundaries of the parameter space. If the true latent distribution is, for example, degenerate, then the conditional MLE is still best asymptotically normal, but there may be estimators, such as the mixture MLE, that will gain superior performance at the sacrifice of asymptotic normality.

8.2.5. *Equivalence theorem for mixture MLE.* If the conditional MLE is fully efficient in the mixture setting, then one must ask "what about the semiparametric mixture MLE?"

Since logic suggests that the MLE for the specified model should be efficient therein, this suggests that perhaps the mixture MLE for the parameters of interest will be asymptotically equivalent to the conditional MLEs.

We are now in a position to address this question using our equivalence theorem. We return to that setting with $\theta = \beta$, $Q = G$, and L_J, L_c and L_m are the mixture likelihood, the conditional likelihood and the marginal likelihood of S, respectively. We check our conditions for equivalence of estimation. Recall that for each (β, G) we need to find G^* with certain properties. The first,

$$L_c(\beta, G) = L_c(\beta, G^*),$$

is trivial here because the conditional likelihood does not depend on G. Thus all we need to do is to find G^* satisfying

$$p_j(\beta, G^*) = \frac{n_j}{n} \qquad \forall j, \tag{8.3}$$

where $p_j(\beta, G) = \Pr[S = j]$, the marginal distribution of the sufficient statistic S.

We now summarize the extensive treatment of this problem in Lindsay, Clogg and Grego (1991).

We have already developed the tools in Chapter 2 to study questions of existence such as posed by (8.3). The subject at hand is the mixture density under fixed β of the statistic S, and so we can plot the unicomponent models as a curve $\{\mathbf{p}(\beta, \phi) \colon \phi \in \Omega\}$ in the simplex, form the convex hull of mixture densities and then ask the question: Does the vector of observed proportions $\mathbf{d}$, with coordinates n_j/n, lie inside this mixture set? If it does, then there exists a solution G^* to (8.3); otherwise, there does not.

Thus, in fact, the answer depends on the observed empirical distribution of the variable S. If it is equal to a mixture density vector, then the conditional and mixture maximum likelihood estimators of the item parameters are *identical*, and otherwise not necessarily. It is important from a conceptual point of view, however, that when the model is correct, then the estimators are identical, with probability 1, for a large enough sample size.

For completeness we add that this asymptotic result relies on the distribution G having enough support points that the vector of true probabilities $\Pr[S{=}j]$ does not lie on the boundary of the mixture set. If this occurs, the semiparametric MLE of θ is not asymptotically equivalent to the conditional MLE, nor even normally distributed. This question can be addressed quite explicitly in the two-item model. The only boundary situation here is when the true latent distribution has a single point of support. In this case the asymptotic distribution of the maximum likelihood estimator of θ is that of a preliminary test estimator in which one chooses between the conditional estimator and the unicomponent estimator based on a preliminary test of heterogeneity.

In all the data sets examined by Lindsay, Clogg and Grego (1991), the conditional and mixture approaches resulted in the same estimator. Among the other issues addressed therein were algorithms and the identifiability of parameters in the mixture model. The authors note that a key advantage of the mixture approach is that it leads quite naturally to empirical Bayes assessment of a subject's latent parameter value.

8.3. Errors-in-variables and case-control studies. As a second example of the use of the equivalency theorem, we consider the problem of estimating logistic regression parameters in a case-control sampling framework. The presentation here represents the pertinent part of a study by Roeder, Carroll and Lindsay (1993). We will first replicate a famous result of Prentice and Pyke by putting it into the framework of the equivalency theorem, and then we extend it into the case of a measurement error model.

8.3.1. *The joint sampling model.* We start with a sample of data (D_i, X_i) in which D_i is a binary response variable, such as diseased (1) and nondiseased (0), and X_i is a vector of potential explanatory variables. We model the sample with a prospective logistic regression that specifies a parametric conditional distribution of Y given X:

$$\Pr[Y = 1 \mid X = x] = \frac{\exp(\alpha + x'\beta)}{1 + \exp(\alpha + x'\beta)}.$$

The second part of the model, the marginal distribution of X, denoted G, is left completely unspecified. We will assume it is discrete, with density $g(x)$. If we do this, the joint likelihood splits apart into two terms, one depending on α and β alone and the other on G alone:

$$L_J(\alpha, \beta, G) = \prod_{i=1}^{n} \Pr[D = d_i \mid X = x_i; \alpha, \beta] \prod_{i=1}^{n} g(x_i).$$

Thus joint maximum likelihood is easy. The nonparametric MLE of G is the empirical CDF of the x data, and α and β are estimated from the prospective logistic regression model.

8.3.2. *The retrospective model.* We next suppose that the sampling was not carried out randomly, but rather we took a sample from each of the two populations: the *cases* ($D = 1$) and the *controls* ($D = 0$). Let n_1 and n_0 be the size of the samples from the two populations and let $n = n_1 + n_0$. Then, in truth we are sampling from the conditional densities $\Pr[X = x \mid D = d]$ under two different values of d, and the likelihood we should maximize is the conditional likelihood:

$$L_c(\alpha, \beta, G) = \prod \Pr[X = x_i \mid D = d_i].$$

The corresponding marginal likelihood for the data is

$$L_m(\alpha, \beta, G) = \Pr[D = 1]^{n_1} \Pr[D = 0]^{n_0}.$$

8.3.3. *Prentice and Pyke's equivalency.* Prentice and Pyke (1979) established that one could obtain the parameters $\hat{\beta}_c$ that maximized the retrospective likelihood L_c by maximizing the joint likelihood L_J, which amounts to maximizing the prospective logistic regression over β and α. Moreover, the solutions obtained this way satisfied

$$\Pr[D = 1; \hat{\alpha}, \hat{\beta}, \hat{G}] = \frac{n_1}{n}.$$

This last result is a clue that this result is an application of the equivalency theorem in Section 8.2.

To apply the theorem, we must establish that for each set of parameters (α, β, G), there exists another set (α^*, β, G^*) such that the retrospective distributions are the same,

$$\Pr[X = x \mid D = d; \alpha, \beta, G] = \Pr[X = x \mid D = d; \alpha^*, \beta, G^*],$$

but the marginal disease distribution equals the observed proportions exactly:

$$\Pr[D = 1; \alpha^*, \beta, G^*] = \frac{n_1}{n}. \tag{8.4}$$

Let $\Pr[D = 1; \alpha, \beta, G] = p$. For the proof of this claim the reader should check that

$$\alpha^* = \alpha + \log[n_1(1 - p)/pn_0]$$

and

$$dG^*(x) \propto \frac{[1+\exp(\alpha^*+\beta x)]}{[1+\exp(\alpha+\beta x)]} dG(x)$$

do the trick. One key point here is that unlike our earlier application, we can satisfy (8.4) for *every* case-control sampling fraction n_1/n that is not zero. The Prentice–Pyke result follows directly.

8.3.4. *The measurement error extension.* The extension we desire to make incorporates one further level of difficulty. Suppose that the desired regressor variable is measured with error. That is, instead of measuring X directly, we observe the surrogate variable $S_i = X_i + \text{error}$. More precisely, we suppose knowledge of a density

$$f(s|x) = \Pr[S = s \mid X = x],$$

possibly with unknown parameters, for the distribution of S_i given $X_i = x$. In our terms, the variable X_i is the latent variable, and what we desire is the latent logistic regression of y on x. The mechanism generating the errors is assumed to be independent of the regression of interest, so that

$$\Pr[D = d \mid S = s, X = x] = \Pr[D = d \mid X = x].$$

The joint distribution of the observables therefore has the mixture form

$$\Pr[D = d, S = s] = \int \Pr[D = d \mid X = x] \Pr[S = s \mid X = x]\, dG(x). \tag{8.5}$$

8.3.5. *The extended equivalency result.* The question now arises: Can we simply maximize the *joint* likelihood

$$L_J(\alpha, \beta, G) = \prod \Pr[D = d_i, S = s_i],$$

which from (8.5) has a standard mixture form, when we want to find the regression parameters β that maximize the *retrospective* likelihood:

$$L_c(\alpha, \beta, G) = \prod \Pr[S = s_i \mid D = d_i]?$$

To maximize the latter, we would have to maximize over a nonstandard likelihood containing ratios of mixture probabilities.

The answer follows directly from the identifiability result we used in the no-measurement-error example. If (α, β, G) are the original parameters, then let (α^*, β, G^*) be as in the previous result. These parameters still give a distribution for D that fits the proportions of cases and controls, because the marginal distribution of D does not depend on the measurement of S. Moreover, if these parameters give identical distributions for $X|D$, then they must give identical distributions for $S|D$.

Furthermore, when we maximize L_J we achieve a perfect fit of the marginal proportions of cases and controls, as in the Prentice–Pyke result. We note that it is critical to this result that G be modeled nonparametrically, so that the equivalency equations (8.1) and (8.2) have a solution.

Roeder, Carroll and Lindsay (1993) have an extended version of this problem in which both X and S are measured on a subsample, as well as considerably more on the practical problem and the efficiency of the maximum likelihood procedure.

8.4. A mixture index of fit. As a last semiparametric example, we will introduce a very different kind of model that arises from a mixture point of view. It provides an interesting extension of some of the techniques we have discussed. The original source for this analysis is Rudas, Clogg and Lindsay (1994).

8.4.1. *The problem.* We suppose that there exists a simple *baseline* probability model $f(x;\beta)$ for the data X that we wish to use to make inferences for a population. It might, for example, be the model of independence for a multiway contingency table. However, we know that it cannot be a perfectly correct description of the population. We would like a simple interpretable measure of how close the baseline model is to being correct—one that we could easily use to evaluate the predictive capabilities of the baseline model. As an illustration, consider the data in Table 8.1, taken from Diaconis and Efron (1985).

We have a simple cross-classification of variables. The chi-squared goodness-of-fit for the independence model gives us a test statistic of 138.29 on 9 degrees of freedom, so we certainly reject that model. However, from this information we cannot tell if it is a reasonably close description of the data anyway, and our rejection arose because we have a relatively large sample size.

8.4.2. *The concept.* We let $g(x)$ be the true distribution of the data. We suppose that it can be written as a mixture of a baseline model density and a second completely arbitrary density $q(x)$:

$$g(x) = (1-\pi)f(x;\beta) + \pi q(x). \tag{8.6}$$

Note that the density $q(x)$ represents a *lack-of-fit* or *outlier* distribution. [Instead of letting $q(x)$ be arbitrary, we could here specify that $q(x)$ is an element of some large class of densities that includes all the model densities and that is also sure to include the true density.]

TABLE 8.1
Cross-classification of eye color and hair color

	Hair color			
Eye color	**Black**	**Brunette**	**Red**	**Blonde**
Brown	68	119	26	7
Blue	20	84	17	94
Hazel	15	54	14	10
Green	5	29	14	16

The variables π and $q(x)$ in the mixture representation (8.6) are not unique, because once a representation is given using π and q, one can construct others by the action of "moving some of the baseline model into the lack-of-fit density" as follows:

$$(8.7)\quad g(x) = (1-\pi-\varepsilon)f(x;\beta) + (\pi+\varepsilon)\left[\frac{\pi}{\pi+\varepsilon}q(x) + \frac{\varepsilon}{\pi+\varepsilon}f(x;\beta)\right],$$

provided that ε is sufficiently small that the new mixture weight $\pi+\varepsilon$ is in [0, 1]. However, one can turn the parameter π into something well defined and interpretable by letting π^*, the *lack-of-fit index* for the density g, be the *smallest* π one could use in such a representation:

$$\pi^*(\mathbf{g}) = \inf\{\pi: g(x) = (1-\pi)f(x;\beta) + \pi q(x)\}.$$

We will assume that the class of baseline models is closed, so that there exists a representation of the form

$$g(x) = (1-\pi^*)f(x;\beta^*) + \pi^* q^*(x).$$

The parameter π^* has a simple interpretation: It is the smallest fraction of the population that must be removed before the baseline model would fit perfectly. Additionally, we can interpret $(1-\pi^*)$ as the maximal fraction of the population that can be described exactly by the baseline model, and so the population fraction to which the model-based inference applies.

8.4.3. *Application to the multinomial.* This modeling scheme leads directly to a method of estimation if we are in a contingency table setting. We continue as before, but substitute the variable t for x to remind us that it is an index for the cells of a multinomial contingency table. We also recall that in this case, we have the nonparametric multinomial MLE, which is just $d(t)$, the observed cell proportions. Hence we can use $\pi^*(\mathbf{d})$ to estimate $\pi^*(\mathbf{g})$.

To better understand how this works, we first consider all multinomial densities $p(t)$ such that their lack-of-fit $\pi^*(\mathbf{p})$ is less than some fixed value π_0:

$$\mathscr{H}_{\pi_0}(\text{model}) = \{p(t): \pi^*(\mathbf{p}) \le \pi_0\}.$$

We can think of $\mathscr{H}_{\pi_0}(\text{model})$ as representing the set of all multinomial densities whose lack-of-fit is acceptably small. We note that because of the shuffling property (8.7) we can also describe $\mathscr{H}_{\pi_0}(\text{model})$ as being the set of multinomial probability vectors for which there exists *some* π_0 representation:

$$\mathscr{H}_{\pi_0}(\text{model}) = \{p(t) = (1-\pi_0)f(t;\beta) + \pi_0 q(t), \text{for some } \beta \text{ and } q(\cdot)\}.$$

In the simplest case, where the baseline model is just a single known distribution $f(t)$, this is just a simplex of the form

$$\mathscr{H}_{\pi_0}(\mathbf{f}) = \text{conv}\{(1-\pi_0)\mathbf{f} + \pi_0\mathbf{e}_t: t = 1,\ldots,T\}.$$

We invite the reader to verify this statement. [The key here is that an arbitrary multinomial distribution $q(t)$ can be written as $\mathbf{q} = \sum q(t)\mathbf{e}_t$.]

The next step is to note that we can find $\mathscr{H}_{\pi_0}$(model) simply by taking the union of the simplices generated by individual baseline model elements:

$$\mathscr{H}_{\pi_0}(\text{model}) = \bigcup_{\beta} \mathscr{H}_{\pi_0}(\mathbf{f}_\beta).$$

It follows that $\mathscr{H}_{\pi_0}$(model) is itself not necessarily a convex set. However, it is true that as we increase π, say from π_0 to π_1, we increase the set of acceptable models:

$$\mathscr{H}_{\pi_0}(\text{model}) \subset \mathscr{H}_{\pi_1}(\text{model}). \tag{8.8}$$

We also note that $\mathscr{H}_0$(model) is just the family of baseline models and that $\mathscr{H}_1$(model) is the full set of multinomial models.

8.4.4. *Maximum likelihood estimation.* This said, we can find the maximum likelihood estimator of the true multinomial density $g(t)$ under the constraint that $\mathscr{H}_{\pi_0}$(model) is our class of acceptable models. We can think of this as estimating the baseline model $f(t;\beta)$, but allowing for a contamination fraction of up to π_0 from some other distribution. Maximum likelihood on this set is fairly easy, since we can hold π_0 *fixed* and maximize over β and $q(\cdot)$ from the class of all models of the form $(1-\pi_0)f(t;\beta)+\pi_0 q(t)$.

Rudas, Clogg and Lindsay (1994) describe an EM algorithm approach that works, but is quite slow. Xi and Lindsay (1995) give a more efficient sequential quadratic programming method.

If we let the *profile* log likelihood $L^*(\pi)$ be the value of the log likelihood after we maximize over $\mathscr{H}_{\pi_0}$, then it is clear from the nesting of the models (8.8) that $L^*(\pi)$ is increasing in π. Let $\hat{L}$ be the likelihood of $\mathbf{d}$, the nonparametric multinomial MLE. We can visualize our acceptable model sets $\mathscr{H}_\pi$ growing in π until just as $\hat{\pi}^* = \pi^*(\mathbf{d})$, we find that $\mathbf{d}$ is in the boundary of $\mathscr{H}_{\hat{\pi}^*}$. From that value of π on, $\mathbf{d}$ is the maximum likelihood estimator from the model set $\mathscr{H}_\pi$, and so $L^*(\pi)=\hat{L}$. We can construct a natural measure of the adequacy of a value of π_0 by the likelihood ratio statistic

$$\text{lrs}(\pi_0) = 2\left[\ln \hat{L} - \ln L^*(\pi_0)\right].$$

This statistic becomes zero when π_0 is sufficiently large.

An analysis of the data in the Table 8.1 can be carried out in this manner, and we arrive at the information in Table 8.2. We can see that about 30% of the data would have to be discarded for the independence model to fit the data exactly. The column labeled X^2 gives the corresponding Pearson chi-squared values. The statistic $\hat{\pi}^*_L$ in the table is our next topic.

8.4.5. *Inference on the lack-of-fit index.* One inferential problem that we face is that the parameter estimator $\hat{\pi}^*$ is biased upward. This is most evident when the baseline model is correct, because the true value of π^* is zero, but the sampled values are necessarily positive. Fortunately, when the true value of π_0 is *not* zero, we can use asymptotics to construct a lower confidence limit for it, and so rescue some assurance that a large value of $\hat{\pi}^*$ is truly atypical.

TABLE 8.2
Fit statistics for the semiparametric mixture model applied to the data in Table 8.1

	X^2	$\mathbf{lrs}(\pi)$
0.00	138.29	146.44
0.10	47.35	48.67
0.15	23.74	24.36
0.20	8.55	8.75
0.236 ($= \hat{\pi}^*_L$)	2.57	2.66
0.25	1.38	1.44
0.26	0.83	0.87
0.27	0.42	0.43
0.28	0.16	0.16
0.29	0.02	0.02
0.298 ($= \hat{\pi}^*$)	0.00	0.00
$\pi \geq 0.298$	0.00	0.00

We therefore consider using the likelihood ratio test statistic $\text{lrs}(\pi_0)$ to test the hypothesis $\pi^*(\mathbf{g}) \leq \pi_0$ against $\pi^*(\mathbf{g}) > \pi_0$.

The asymptotics of the situation can be handled by the techniques described in Chapter 4. When π_0 is correct, the null density vector $\mathbf{g}$ sits on the boundary of the set $\mathscr{H}_{\pi_0}$. This set, from its earlier description, is full dimensional within the simplex.

To aid the geometric imagination, we first consider the simple case where the baseline model is a single density $\mathbf{f}$ and $\mathscr{H}_{\pi_0}$(model) is just the simplex $\mathscr{H}_{\pi_0}(\mathbf{f})$ containing $\mathbf{f}$. After transforming to the dagger space, the simplex is still a simplex, and as long as $\mathbf{g}$ is on one of the flat full-dimensional faces of this simplex, then the model cone generated by the models in $\mathscr{H}_{\pi_0}(\mathbf{f})$ are an entire half space. All directions that one can move while staying in the simplex face correspond to "nuisance score" directions, and any vector which is E_0 orthogonal to the face corresponds to the corrected score for π. Since if we go in one direction we go into the model and the other way we leave it, we are in a setting where the asymptotic distribution of likelihood ratio test statistic for $\pi^*(\mathbf{g}) \leq \pi_0$ versus $>$ has the limiting distribution $0.5\chi_0^2 + 0.5\chi_1^2$.

Of course, when the true distribution is not on a face of the simplex $\mathscr{H}_{\pi_0}(\mathbf{f})$, but on an edge or corner or other lower dimensional surface, the analysis becomes considerably more difficult. The geometry shows that the test statistic will have a chi-bar-squared distribution that is stochastically larger than $0.5\chi_0^2 + 0.5\chi_1^2$. Unfortunately, since this makes us more likely to reject if the true distribution is such a boundary point and therefore if one uses the distribution $0.5\chi_0^2 + 0.5\chi_1^2$, then the test procedure is anticonservative. However, we do note that the region of parameter values where the test based on $0.5\chi_0^2 + 0.5\chi_1^2$ is anticonservative has Lebesgue measure zero and might be presumed to have prior probability of zero.

These arguments extend beyond the single distribution model $\mathscr{H}_{\pi_0}(\mathbf{f})$ to the general case $\mathscr{H}_{\pi_0}$(model) by using the fact that for a general baseline model, the π-model set is a union of simplices and so has an open interior in the neighborhood of each boundary point. If the surface is smooth at the null hypothesis density, then the half-space conal geometry described above still holds true. However, the problem with "edges" is now more difficult to analyze, and one could lose the convex cone structure that guaranteed that using $0.5\chi_0^2 + 0.5\chi_1^2$ would be anticonservative.

Despite these technical difficulties, we think a reasonable procedure is the use of the distribution $0.5\chi_0^2 + 0.5\chi_1^2$ as a guide for constructing tests and confidence intervals. Inverting this test gives an upper confidence interval of the form

$$\{\pi : \text{lrs}(\pi) \leq \chi_1^2(2\alpha)\} = [\pi_L^*, 1].$$

If we examine Table 8.2, we see that the 95% lower limit for π^* in this data is about 24%, still a rather large lack-of-fit fraction.

Bibliography

AITKIN, M. and RUBIN, D. B. (1985). Estimation and hypothesis testing in finite mixture models. *J. Roy. Statist. Soc. Ser. B* **47** 67–75.

AITKIN, M. and WILSON, G. T. (1980). Mixture models, outliers, and the EM algorithm. *Technometrics* **22** 325–331.

ANDERSON, E. B. (1973). *Conditional Inference and Models for Measuring*. Mentalhygiejnisk Forlag, Copenhagen.

BERAN and HALL (1992). Estimating coefficient distributions in random coefficient regressions. *Ann. Statist.* **20** 1970–1984.

BICKEL, P. J., KLAASSEN, C. J., RITOV, Y. and WELLNER, J. A. (1993). *Efficient and Adaptive Estimation for Semiparametric Models*. Johns Hopkins Univ. Press.

BÖHNING, D. (1985). Numerical estimation of a probability measure. *J. Statist. Plann. Inference* **11** 57–69.

BÖHNING, D. (1995). A review of reliable maximum likelihood algorithms for semiparametric mixture models. *J. Statist. Plann. Inference*. To appear.

BÖHNING, D., DIETZ, E., SCHAUB, R., SCHLATTMAN, P. and LINDSAY, B. G. (1994). The distribution of the likelihood ratio for mixtures of densities from the one parameter exponential family. *Ann. Inst. Statist. Math.* **46** 373–388.

BÖHNING, D., SCHLATTMAN, P. and LINDSAY, B. G. (1992). Computer assisted analysis of mixtures (C.A.MAN): Statistical algorithms. *Biometrics* **48** 283–304.

BRÄNNÄS, K. and ROSENQVIST, G. (1994). Semiparametric estimation of heterogeneous count data models. *European J. Oper. Res.* **76** 247–258.

BUTLER, S. M. and LOUIS, T. A. (1992). Random effects models with nonparametric priors. *Statist. Med.* **11** 1981–2000.

CHEN, J. (1993). A new approach to testing the number of components in finite mixture models. Unpublished manuscript, Univ. Waterloo.

CHERNOFF, H. (1954). On the distribution of the likelihood ratio. *Ann. Math. Statist.* **25** 573–578.

CLOGG, C. C. (1995). Latent class models. In *Handbook of Statistical Modeling for the Social and Behavioral Sciences* (G. Arminger, C. C. Clogg and M. E. Sobel, eds.) 311–359. Plenum, New York.

DALAL, S. R. and HALL, W. J. (1983). Approximating priors by mixtures of natural conjugate priors. *J. Roy. Statist. Soc. Ser. B*, **45** 278–286.

DAVIDIAN, M. and GALLANT, A. R. (1992). Smooth nonparametric maximum likelihood estimation for population pharmacokinetics, with application to quinidine. *Journal of Pharmacokinetics and Biopharmaceutics* **20** 529–556.

DAVIES, R. B. (1993). Nonparametric control for residual heterogeneity in modeling recurrent behavior. *Comput. Statist. Data Anal.* **16** 143–160.

DEMPSTER, A. P., LAIRD, N. M. and RUBIN, D. B. (1977). Maximum likelihood from incomplete data via the EM algorithm. *J. Roy. Statist. Soc. Ser. B* **39** 1–22.

DERSIMONIAN, R. (1986). Maximum likelihood estimation of a mixing distribution. *J. Roy. Statist. Soc. Ser. C* **35** 302–309.

DERSIMONIAN, R. (1989). Tests of homogeneity in mixed models. *Technical Report*. Dept. Public Health, School of Medicine, Yale University.

DERSIMONIAN, R. (1990). Correction to algorithm AS 221: maximum likelihood estimation of a mixing distribution. *J. Roy. Statist. Soc. Ser. C* **39** 176.

DIACONIS, P. (1977). Finite forms of de Finetti's theorem on exchangeability. *Synthese* **36** 271–281.

DIACONIS, P. and EFRON, B. (1985). Testing for independence in a two-way table: New interpretations of the chi-square statistic. *Ann. Statist.* **13** 845–874.

DIACONIS, P. and YLVISAKER, D. (1979). Conjugate priors for exponential families. *Ann. Statist.* **7** 269–281.

DO, K. and MCLACHLAN, G. J. (1984). Estimation of mixing proportions: A case study. *J. Roy. Statist. Soc. Ser. C* **33** 134–140.

EFRON, B. (1986). Double exponential families and their use in generalized linear regression. *J. Amer. Statist. Assoc.* **81** 709–721.

EFRON, B. and MORRIS, C. (1977). Stein's paradox in statistics. *Scientific American* **236** 119–127.

EVERITT, B. S. and HAND, D. J. (1981). *Finite Mixture Distributions*. Chapman and Hall, New York.

EZZET, F. L. and DAVIES, R. B. (1988). A manual for MIXTURE. Center for Applied Statistics, Univ. Lancaster, England.

FAN, J. Q. (1991). On the optimal rates of convergence for nonparametric deconvolution problems. *Ann. Statist.* **19** 1257–1272.

FELLER, W. (1971). *An Introduction to Probability Theory and Its Applications* **2**. Wiley, New York.

FINCH, S. J., MENDELL, N. R. and THODE, H. C. (1989). Probabilistic measures of adequacy of a numerical search for a global maximum. *J. Amer. Statist. Assoc.* **84** 1020–1023.

FISHER, R. A. (1925). *Statistical Methods for Research Workers.*, 1st ed. Oliver and Boyd, Edinburgh.

FOLLMAN, D. A. and LAMBERT, D. (1989). Generalizing logistic regression by nonparametric mixing. *J. Amer. Statist. Assoc.* **84** 295–300.

FOLLMAN, D. A. and LAMBERT, D. (1991). Identifiability of finite mixtures of logistic regression models. *J. Statist. Plann. Inference* **27** 375–381.

FRASER, D. A. S. and MASSAM, H. (1989). A mixed primal-dual bases algorithm for regression under inequality constraints. *Scand. J. Statist.* **16** 65–74.

FURMAN, D. and LINDSAY, B. G. (1994a). Measuring the relative effectiveness of moment estimators as starting values in maximizing mixture likelihoods. *Comput. Statist. Data Anal.* **17** 493–507.

FURMAN, D. and LINDSAY, B. G. (1994b). Testing for the number of components in a mixture of normal distributions using moment estimators. *Comput. Statist. Data Anal.* **17** 473–492.

GALLANT, A. R. and NYCHKA, D. W. (1987). Seminonparametric maximum likelihood estimation. *Econometrica* **55** 363–390.

GEISSLER, A. (1889). Beiträge zur Frage des Geschlechts Verhältnisses der Geborenen. Zeitschrift des K. Sächsischen Statistischen Bureaus.

GELFAND, A. E. and DALAL, S. R. (1990). A note on overdispersed exponential families. *Biometrika* **77** 55–64.

GHOSH, J. K. and SEN, P. K. (1985). On the asymptotic performance of the log likelihood ratio statistic for the mixture model and related results. In *Proceedings of the Berkeley Conference in Honor of Jerzy Neyman and Jack Kiefer* (L. M. Le Cam and R. A. Olshen, eds.) **2** 789–806. Wadsworth, Belmont, CA.

GOFFINET, B., LOISEL, P. and LAURENT, B. (1992). Testing in normal mixture models when the proportions are known. *Biometrika* **79** 842–846.

GROENEBOOM, P. and WELLNER, J. A. (1992). Information bounds and nonparametric maximum likelihood estimation. *DMV Sem.* **19**. Birkhäuser, Boston.

HARTIGAN, J. A. (1985). A failure of likelihood asymptotics for normal mixtures. In *Proceedings of the Berkeley Conference in Honor of Jerzy Neyman and Jack Kiefer* (L. M. Le Cam and R. A. Olshen, eds.) **2** 807–810. Wadsworth, Belmont, CA.

HATHAWAY, R. (1985). A constrained formulation of maximum-likelihood estimation for normal mixture distributions. *Ann. Statist.* **13** 795–800.

HECKMAN, J. J. and SINGER, B. (1984). Econometric duration analysis. *J. Econometrics* **24** 63–132.

HEINEN, T. (1993). *Discrete Latent Variable Models*. Tilburg Univ. Press, The Netherlands.

HOEFFDING, W. (1956). One the distribution of the number of successes in independent trials. *Ann. Math. Statist.* **27** 713–721.

HOTELLING, H. (1939). Tubes and spheres in n-spaces, and a class of statistical problems. *Amer. J. Math.* **61** 440–460.

HSI, H., LINDSAY, B. G. and LYNCH, J. (1992). On mixtures of hazards: Nonparametric maximum likelihood in a competing risk model. *Journal of Nonparametric Statistics* **2** 89–103.

JEWELL, N. P. (1982). Mixtures of exponential distributions. *Ann. Statist.* **10** 479–484.

JEWELL, N. P., MALANI, H. M. and VITTINGHOFF, E. (1994). Nonparametric estimation for a form of doubly censored data with application to two problems in AIDS. *J. Amer. Statist. Assoc.* **89** 7–18.

KARLIN, S. (1968). *Total Positivity*. Stanford Univ. Press.

KARLIN, S. and STUDDEN, W. J. (1966). *Tchebysheff Systems: With Applications to Analysis and Statistics*. Wiley, New York.

KIEFER, J. and WOLFOWITZ, J. (1956). Consistency of the maximum likelihood estimator in the presence of infinitely many incidental parameters. *Ann. Math. Statist.* **27** 886–906.

KIEFER, N. M. (1978). Discrete parameter variation: Efficient estimation of a switching regression model. *Econometrica* **46** 427–434.

LAIRD, N. M. (1978). Nonparametric maximum likelihood estimation of a mixing distribution. *J. Amer. Statist. Assoc.* **73** 805–811.

LAIRD, N. M. and LOUIS, T. A. (1991). Smoothing the nonparametric estimate of a prior distribution by roughening. *Comput. Statist. Data Anal.* **12** 27–37.

LAMBERT, D. and ROEDER, K. (1995). Overdispersion diagnostics for generalized linear models. *J. Amer. Statist. Assoc.* To appear.

LESPERANCE, M. L. and KALBFLEISCH, J. D. (1992). An algorithm for computing the nonparametric MLE of a mixing distribution. *J. Amer. Statist. Assoc.* **87** 120–126.

LIANG, K. (1984). The asymptotic efficiency of conditional likelihood methods. *Biometrika* **71** 305–313.

LINDSAY, B. G. (1981). Properties of the maximum likelihood estimator of a mixing distribution. In *Statistical Distributions in Scientific Work* (G. P. Patil, ed.) **5** 95–109. Reidel, Boston.

LINDSAY, B. G. (1983a). The geometry of mixture likelihoods: A general theory. *Ann. Statist.* **11** 86–94.

LINDSAY, B. G. (1983b). Efficiency of the conditional score in a mixture setting. *Ann. Statist.* **11** 486–497.

LINDSAY, B. G. (1983c). The geometry of mixture likelihoods, Part II: The exponential family. *Ann. Statist.* **11** 783–792.

LINDSAY, B. G. (1986). Exponential family mixture models (with least squares estimators). *Ann. Statist.* **14** 124–137.

LINDSAY, B. G. (1989a). On the determinants of moment matrices. *Ann. Statist.* **17** 711–721.

LINDSAY, B. G. (1989b). Moment matrices: Applications in mixtures. *Ann. Statist.* **17** 722–740.

LINDSAY, B. G. and BASAK, P. (1993). Multivariate normal mixtures: A fast consistent method of moments. *J. Amer. Statist. Assoc.* **88** 468–476.

LINDSAY, B. G., CLOGG, C. C. and GREGO, J. (1991). Semi-parametric estimation in the Rasch model and related exponential response models, including a simple latent class model for item analysis. *J. Amer. Statist. Assoc.* **86** 96–107.

LINDSAY, B. G. and LESPERANCE, M. L. (1995). A review of semiparametric mixture models. *J. Statist. Plann. Inference* To appear.

LINDSAY, B. G. and ROEDER, K. (1992a). Residual diagnostics in the mixture model. *J. Amer. Statist. Assoc.* **87** 785–795.

LINDSAY, B. G. and ROEDER, K. (1992b). Moment-based oscillation properties of mixture models. Technical Report **92-3**, Center for Likelihood Studies, Pennsylvania State University.

LINDSAY, B. G. and ROEDER, K. (1993). Uniqueness and identifiability in nonparametric mixtures. *Canad. J. Statist.* **21** 139–147.

LOUIS, T. A. (1982). Finding the observed information matrix when using the EM algorithm. *J. Roy. Statist. Soc. Ser. B* **44** 226–233.

MACDONALD, P. D. M. and PITCHER, T. J. (1979). Age-groups from size-frequency data: A versatile and efficient method of analyzing distribution mixtures. *Journal of the Fisheries Board of Canada* **36** 987–1001.

MALLET, A. (1986). A maximum likeihood estimation method for random coefficient regression models. *Biometrika* **73** 645–656.

MALLET, A., MENTRE, F., STEIMER, J-L. and LOKIEC, F. (1988). Nonparametric maximum likelihood estimation for population pharmacokinetics, with application to cyclosporine. *Journal of Pharmacokinetics and Biopharmaceutics* **16** 311–327.

MARITZ, J. S. and LWIN, T. (1989). *Empirical Bayes Methods*, 2nd ed. Chapman and Hall, London.

MCLACHLAN, G. J. and BASFORD, K. E. (1988). *Mixture Models: Inference and Applications to Clustering*. Dekker, New York.

MILLAR, R. B. (1987). Maximum likelihood estimation of mixed stock fishery composition. *Canadian Journal of Fisheries and Aquatic Sciences* **44** 583–590.

MORAN, P. A. P. (1973). Asymptotic properties of homogeneity tests. *Biometrika* **60** 79–85.

NAIMAN, D. Q. (1986). Conservative confidence bands in curvilinear regression. *Ann. Statist.* **14** 896–906.

NEUHAUS, J. M., HAUCK, W. W. and KALBFLEISCH, J. D. (1992). The effects of mixture distribution misspecification when fitting mixed-effects logistic models. *Biometrika* **79** 755–762.

NEUHAUS, J. M., KALBFLEISCH, J. D. and HAUCK, W. W. (1991). A comparison of cluster-specific and population-averaged approaches for analyzing correlated binary data. *Internat: Statist. Rev.* **59** 25–35.

NEYMAN, J. (1959). Optimal asymptotic tests of composite statistical hypotheses. In *Probability and Statistics. The Harald Cramér Volume* (U. Grenander, ed.) 213–234. Wiley, New York.

NEYMAN, J. and SCOTT, E. L. (1948). Consistent estimates based on partially consistent observations. *Econometrica* **16** 1–32.

NEYMAN, J. and SCOTT, E. L. (1966). On the use of $C(\alpha)$ optimal tests of composite hypotheses. *Bull. Inst. Int. Statist.* **41 I** 477–97.

OWEN, A. B. (1988). Empirical likelihood ratio confidence intervals for a single functional. *Biometrika* **75** 237–249.

PÓLYA, G. and SZEGŐ, G. (1925). *Aufgaben und Lehrsätze aus der Analysis* **2**. Springer, Berlin.

PRAKASA RAO, B. L. S. (1992). *Identifiability in Stochastic Models, Characterization of Probability Distributions*. Academic, Boston.

PRENTICE, R. L. and PYKE, R. (1979). Logistic disease incidence models and case-control studies. *Biometrika* **66** 403–412.

QUINN, B. G., MCLACHLAN, G. J. and HJORT, N. L. (1987). A note on the Aitkin–Rubin approach to hypothesis testing in mixture models. *J. Roy. Statist. Soc. Ser. B* **49** 311–314.

REDNER, R. and WALKER, H. F. (1984). Mixture densities, maximum likelihood and the EM algorithm. *SIAM Rev.* **26** 195–239.

ROBBINS, H. (1950). A generalization of the method of maximum likelihood: Estimating a mixing distribution (abstract). *Ann. Math. Statist.* **21** 314–315.

ROBERTS, A. W. and VARBERG, D. E. (1973). *Convex Functions*. Academic, New York.

ROBERTSON, C. A. and FRYAR, J. G. (1969). Some descriptive properties of normal mixtures. *Skand. Aktuarier Tidskr.* **52** 137–146.

ROBERTSON, T., WRIGHT, F. T. and DYKSTRA, R. L. (1986). *Order Restricted Statistical Inference*. Wiley, Chichester.

ROEDER, K. (1990). Density estimation with confidence sets exemplified by superclusters and voids in the galaxies. *J. Amer. Statist. Assoc.* **85** 617–624.

ROEDER, K. (1994). A graphical technique for determining the number of components in a mixture of normals. *J. Amer. Statist. Assoc.* **89** 487–495.

ROEDER, K., CARROLL, R. J. and LINDSAY, B. G. (1993). A nonparametric mixture approach to case-control studies with errors in covariables. Technical Report 94-1, Center for Likelihood Studies, Pennsylvania State Univ.

ROEDER, K., DEVLIN, B. and LINDSAY, B. G. (1989). Application of maximum likelihood methods to population genetic data for the estimation of individual fertilities. *Biometrics* **45** 363–379.

RUDAS, T., CLOGG, C. C. and LINDSAY, B. G. (1994). A new index of fit based on mixture methods for the analysis of contingency tables. *J. Roy. Statist. Soc. Ser. B* **56** 623–639.

SCHOLZ, F. W. (1980). Towards a unified definition of maximum likelihood. *Canad. J. Statist.* **8** 193–203.

SHAKED, M. (1980). On mixtures from exponential families. *J. Roy Statist. Soc. Ser. B* **42** 192–198.

SHAPIRO, A. (1985). Asymptotic distribution of test statistics in the analysis of moment structures under inequality constraints. *Biometrika* **72** 133–144.

SHEPP, L. A. and VARDI, Y. (1982). Maximum likelihood reconstruction in positron emission tomography. *IEEE Transactions on Medical Imaging* **1** 113–122.

SILVEY, S. D. (1980). *Optimal Design*. Chapman and Hall, New York.

SIMAR, L. (1976). Maximum likelihood estimation of a compound Poisson process. *Ann. Statist.* **4** 1200–1209.

TITTERINGTON, D. M., SMITH, A. F. M. and MAKOV, U. E. (1985). *Statistical Analysis of Finite Mixture Distributions*. Wiley, New York.

USPENSKY, J. V. (1937). *Introduction to Mathematical Probability*, 1st ed. McGraw-Hill, New York.

VARDI, Y. and LEE, D. (1993). From image deblurring to optimal investments: Maximum likelihood solutions for positive linear inverse problems. *J. Roy. Statist. Soc. Ser. B* **55** 569–599.

VARDI, Y., SHEPP, L. A. and KAUFMAN, L. (1985). A statistical model for positron emission tomography. *J. Amer. Statist. Assoc.* **80** 8–20.

WALTER, G. G. and HAMEDANI, G. G. (1991). Bayes empirical Bayes estimation for natural exponential families with quadratic variance functions. *Ann. Statist.* **19** 1191–1224.

WING, G. M. and ZAHRT, J. D. (1991). *A Primer on Integral Equations of the First Kind: The Problem of Deconvolution and Unfolding*. SIAM, Philadelphia.